Unifying Themes in Complex Systems

Volume VII

Springer Complexity

Springer Complexity is a publication program, cutting across all traditional disciplines of sciences as well as engineering, economics, medicine, psychology and computer sciences, which is aimed at researchers, students and practitioners working in the field of complex systems. Complex Systems are systems that comprise many interacting parts with the ability to generate a new quality of macroscopic collective behavior through self-organization, e.g., the spontaneous formation of temporal, spatial or functional structures. This recognition, that the collective behavior of the whole system cannot be simply inferred from the understanding of the behavior of the individual components, has led to various new concepts and sophisticated tools of complexity. The main concepts and tools – with sometimes overlapping contents and methodologies – are the theories of self-organization, complex systems, synergetics, dynamical systems, turbulence, catastrophes, instabilities, nonlinearity, stochastic processes, chaos, neural networks, cellular automata, adaptive systems, and genetic algorithms.

The topics treated within Springer Complexity are as diverse as lasers or fluids in physics, machine cutting phenomena of workpieces or electric circuits with feedback in engineering, growth of crystals or pattern formation in chemistry, morphogenesis in biology, brain function in neurology, behavior of stock exchange rates in economics, or the formation of public opinion in sociology. All these seemingly quite different kinds of structure formation have a number of important features and underlying structures in common. These deep structural similarities can be exploited to transfer analytical methods and understanding from one field to another. The Springer Complexity program therefore seeks to foster cross-fertilization between the disciplines and a dialogue between theoreticians and experimentalists for a deeper understanding of the general structure and behavior of complex systems.

The program consists of individual books, books series such as "Springer Series in Synergetics", "Institute of Nonlinear Science", "Physics of Neural Networks", and "Understanding Complex Systems", as well as various journals.

New England Complex Systems Institute

President
Yaneer Bar-Yam
New England Complex Systems Institute
238 Main Street Suite 319
Cambridge, MA 02142, USA

NECSI

For over ten years, the New England Complex Systems Institute (NECSI) has been instrumental in the development of complex systems science and its applications. NECSI conducts research, education, knowledge dissemination, and community development around the world for the promotion of the study of complex systems and its application for the betterment of society.

NECSI was founded by faculty of New England area academic institutions in 1996 to further international research and understanding of complex systems. The science of complex systems is a growing field that aims to understand how parts of a system give rise to the system's collective behaviors, and how it interacts with its environment. These questions can be studied in general, and they are also relevant to all traditional fields of science.

Social systems formed (in part) out of people, the brain formed out of neurons, molecules formed out of atoms, and the weather formed from air flows are all examples of complex systems. The field of complex systems intersects all traditional disciplines of physical, biological and social sciences, as well as engineering, management, and medicine. Advanced education in complex systems attracts professionals, as complex-systems science provides practical approaches to health care, social networks, ethnic violence, marketing, military conflict, education, systems engineering, international development and counter-terrorism.

The study of complex systems is about understanding indirect effects. Problems we find difficult to solve have causes and effects that are not obviously related. Pushing on a complex system "here" often has effects "over there" because the parts are interdependent. This has become more and more apparent in our efforts to solve societal problems or avoid ecological disasters caused by our own actions. The field of complex systems provides a number of sophisticated tools, some of them conceptual, helping us think about these systems; some of them analytical, for studying these systems in greater depth; and some of them computer-based, for describing, modeling or simulating them.

NECSI research develops basic concepts and formal approaches as well as their applications to real-world problems. Contributions of NECSI researchers include studies of networks, agent-based modeling, multiscale analysis and complexity, chaos and predictability, evolution, ecology, biodiversity, altruism, systems biology, cellular response, health care, systems engineering, negotation, military conflict, ethnic violence, and international development.

Throughout the year, NECSI's classes, seminars, conferences and other programs assist students and professionals alike in their understanding of complex systems. Courses have been taught in Australia, Canada, China, Colombia, France, Italy, Japan, Korea, Portugal, Russia and many states of the U.S. NECSI also sponsors postdoctoral fellows, provides research resources, and hosts the International Conference on Complex Systems as well as discussion groups and web resources.

New England Complex Systems Institute
Book Series

Series Editor
Dan Braha
New England Complex Systems Institute
238 Main Street Suite 319
Cambridge, MA 02142, USA

The world around us is full of the wonderful interplay of relationships and emergent behaviors. The beautiful and mysterious way that atoms form biological and social systems inspires us to new efforts in science. As our society becomes more concerned with how people are connected to each other than how they work independently, so science has become interested in the nature of relationships and relatedness. Through relationships elements act together to become systems, and systems achieve function and purpose. The elements act together to become systems, and systems achieve function and purpose. The study of complex systems is remarkable in the closeness of basic ideas and practical implications. Advances in our understanding of complex systems give new opportunities for insight in science and improvement of society. This is manifest in the relevance to engineering, medicine, management and education. We devote this book series to the communication of recent advances and reviews of revolutionary ideas and their application to practical concerns.

Unifying Themes
in Complex Systems VII

Proceedings of the
Seventh International Conference on Complex
Systems

**Edited by Ali Minai, Dan Braha
and Yaneer Bar-Yam**

Ali A. Minai
Univeristy of Cincinnati
Department of Electrical and
Computer Engineering, and Computer Science
P.O. Box 210030, Rhodes Hall 814
Cincinnati, OH 45221-0030, USA
Email: Ali.Minai@uc.edu

Dan Braha
New England Complex Systems Institute
238 Main Street Suite 319
Cambridge, MA 02142, USA
Email: braha@necsi.edu

Yaneer Bar-Yam
New England Complex Systems Institute
238 Main Street Suite 319
Cambridge, MA 02142, USA
Email: yaneer@necsi.edu

This volume is part of the
New England Complex Systems Institute Series on Complexity

ISBN 978-3-642-18002-6 Springer Berlin Heidelberg New York

Springer is a part of Springer Science+Business Media
springer.com
NECSI Cambridge, Massachusetts 2012

CONTENTS — 2007 CONFERENCE

INTRODUCTION

The mysteries of highly complex systems that have puzzled scientists for years are finally beginning to unravel thanks to new analytical and simulation methods. Better understanding of concepts like complexity, emergence, evolution, adaptation and self-organization have shown that seemingly unrelated disciplines have more in common than we thought. These fundamental insights require interdisciplinary collaboration that usually does not occur between academic departments. This was the vision behind the first International Conference on Complex Systems in 1997: not just to present research, but to introduce new perspectives and foster collaborations that would yield research in the future.

As more and more scientists began to realize the importance of exploring the unifying principles that govern all complex systems, the 2007 ICCS attracted a diverse group of participants representing a wide variety of disciplines. Topics ranged from economics to ecology, from physics to psychology and from business to biology. Through pedagogical, breakout and poster sessions, conference attendees shared discoveries that were significant both to their particular field of interest, as well as to the general study of complex systems. These volumes contain the proceedings from that conference.

Even with the seventh ICCS, the science of complex systems is still in its infancy. In order for complex-systems science to fulfill its potential to provide a unifying framework for various disciplines, it is essential to provide a standard set of conventions to facilitate communication. This is another valuable function of the conference: it allowed an opportunity to develop a common foundation and language for the study of complex systems.

These efforts have produced a variety of new analytic and simulation techniques that have proven invaluable in the study of physical, biological and social systems. New methods of statistical analysis led to better understanding of polymer formation and complex fluid dynamics; further development of these methods has deepened our understanding of patterns and networks. The application of simulation techniques such as agent-based models, cellular automata and Monte Carlo calculations to complex systems has increased our ability to understand and even predict behavior of systems which once seemed completely unpredictable.

The concepts and tools of complex systems are of interest not only to scientists, but also to corporate managers, doctors, political scientists and policy makers. The same rules that govern neural networks apply to social or corporate networks, and professionals have started to realize how valuable these concepts are to their individual fields. The International Conferences on Complex Systems have provided the opportunity for professionals to learn the basics of complex systems and share their real-world experience in applying these concepts.

ORGANIZATION

Conference Chair:

Yaneer Bar-Yam - NECSI

Executive Committee:

Dan Braha - University of Massachusetts, Dartmouth
Joel MacAuslan - Speech Technology and Applied Research
Ali Minai - University of Cincinnati
Hiroki Sayama - Binghamton University, SUNY

Logistics:

Sageet Braha
Eric Downes
Nina Duraiswami
Luke Evans
Seth Frey
Debra Gorfine
Konstantin Koupstov
Matt Lieberman
David Pooley
Blake Stacey
Cam Terwilliger
Greg Wolfe
Mark Woolsey

Program Committee:

Yaneer Bar-Yam - NECSI
Philippe Binder - University of Hawaii, Hilo
Dan Braha - University of Massachusetts, Dartmouth
Irene Conrad - Texas A&M, Kingsford
Fred Discenzo - Rockwell Automation
Irina Ezhkova - International Institute of Applied Technologies, Brussels
Philip Fraundorf - University of Missouri, St. Louis
Carlos Gershenson - NECSI
Dion Harmon - NECSI
Mark Kon - Boston University
Joel MacAuslan - Speech Technology and Applied Research
Ed Marcus - Marcus Laboratories
Ali Minai - University of Cincinnati
Lael Parrott - Université de Montréal
Daniel Polani - University of Hertfordshire
David Saakian - Yerevan Physics Institute

Hiroki Sayama - Binghamton University, SUNY
Jeff Schank - University of California, Davis
Sanith Wijesinghe - MillenniumIT

Additional Support:

Birkhäuser
Books and Books
Harvard University Press
Oxford University Press
Random House
Springer

Founding Organizing Committee:

Philip W. Anderson - Princeton University
Kenneth J. Arrow - Stanford University
Michel Baranger - MIT
Per Bak - Niels Bohr Institute
Charles H. Bennett - IBM
William A. Brock - University of Wisconsin
Charles R. Cantor - Boston University
Noam A. Chomsky - MIT
Leon Cooper - Brown University
Daniel Dennett - Tufts University
Irving Epstein - Brandeis University
Michael S. Gazzaniga - Dartmouth College
William Gelbart - Harvard University
Murray Gell-Mann CalTech/Santa Fe Institute
Pierce-Gilles de Gennes - ESPCI
Stephen Grossberg - Boston University
Michael Hammer - Hammer & Co
John Holland - University of Michigan
John Hopfield - Princeton University
Jerome Kagan - Harvard University
Stuart A. Kauffman - Santa Fe Institute
Chris Langton - Santa Fe Institute
Roger Lewin - Harvard University
Richard C. Lewontin - Harvard University
Albert J. Libchaber - Rockefeller University
Seth Lloyd - MIT
Andrew W. Lo - MIT
Daniel W. McShea - Duke University
Marvin Minsky - MIT
Harold J. Morowitz - George Mason University
Alan Perelson - Los Alamos National Lab
Claudio Rebbi - Boston University

Herbert A. Simon - Carnegie-Mellon University
Temple F. Smith - Boston University
H. Eugene Stanley - Boston University
John Sterman - MIT
James H. Stock - Harvard University
Gerald J. Sussman - MIT
Edward O. Wilson - Harvard University
Shuguang Zhang - MIT

Session Chairs:

Philippe Binder - University of Hawaii, Hilo
Dan Braha - University of Massachusetts, Dartmouth
Irene Conrad - Texas A&M, Kingsford
René Doursat - CNRS and Ecole Polytechnique, Paris
Phil Fellman - Southern New Hampshire University School of Business
Carlos Gershenson - NECSI
Anne-Marie Grisogono - DSTO
Helen Harte - NECSI
Alfred Hubler - University of Illinois, Urbana-Champaign
Eric Klopfer - MIT, NECSI
Corey Lofdahl - BAE Systems
Joel MacAuslan - Speech Technology and Applied Research
Gottfried Mayer-Kress - PSU
Ali Minai - University of Cincinnati
David Miron - CSIRO Livestock Industries
Eve Mitleton-Kelly - London School of Economics
Gary Nelson - Homeland Security Institute
Daniel Polani - University of Hertfordshire
Jonathan vos Post - Computer Futures, Inc.
Robert Savit - University of Michigan
Jeff Schank - University of California, Davis
William Sulis - McMaster University
Len Troncale - California State Polytechnic University
Marlene Williamson
Ian Wilkinson - University of New South Wales

CONFERENCE PROGRAMME

SUNDAY, October 28

9:00AM–5:00PM PEDAGOGICAL SESSIONS
ERIC KLOPFER – *Pedagogical Sessions*

- DIANA DABBY – *Musical Complexity*

- ED FREDKIN – *Discrete Universe*

- GYAN BHANOT – *Human Migration*

- LIZ BRADLEY – *Nonlinear Computers*

HIROKI SAYAMA – *Pedagogical Sessions*

- MARTIN BERZINS – *Complex multiscale engineering simulations of fires and explosions*

- EVELYN FOX KELLER – *Evolving Function, Purpose, and Agency*

- FRANNIE LEAUTIER – *The Challenge of Sustainable Development*

- BLAISE AGUERA Y ARCAS – *Multiscale Visualization*

EVENING RECEPTION
BARBARA JASNY (*Science* Magazine) – *Advances in Science and their Communication*

MONDAY, October 29

9:00AM–12:20PM MORNING SESSIONS
ALFRED HUBLER – *Emergence*

- PHILLIP ZIMBARDO – *Human Behavior, Good and Bad*

- SIMON LEVIN – *Ecological and Socioeconomic Systems as Complex Systems*

- JOHN STERMAN – *Understanding Global Warming*

- MARC KIRSHNER – *Plausibility of Life*

2:00PM–5:20PM AFTERNOON SESSIONS
Robert Savit – *Modeling*

- GEOFFREY WEST – *Universal Laws of Biology*

- GREGORY CHAITIN – *Uncomputable Numbers: Turing's 1936 Paper Revisited*

- JOSH EPSTEIN – *Agent-Based Modeling*

- IRVING EPSTEIN – *Spatial Patterns with Global Implications*

6:30PM–9:00PM EVENING SESSIONS
Session Chairs:

- WILLIAM SULIS – *Concepts, Methods and Tools*

- GOTTFRIED MAYER-KRESS – *Non-Linear Dynamics and Pattern Formation*

- JEFF SCHANK – *Evolution and Ecology*

9:00PM–10:00PM POSTER SESSION

TUESDAY, October 30

9:00AM–12:20PM MORNING SESSIONS
ERNEST HARTMANN – *Biological Systems*

- IAIN COUZIN – *Collective Behavior in Animals*

- ANDRE LEVCHENKO – *Complexity from Simplicity: Why Does It Pay to Be Long?*

- HAVA SIEGELMANN – *Computation and Life*

- JANET WILES – *Complex Systems from DNA to Development*

2:00PM–5:00PM AFTERNOON SESSIONS
Session Chairs:

- PHIL FELLMAN – *Aesthetics and Consciousness*

- CARLOS GERSHENSON – *Networks*

- DANIEL POLANI – *Engineering*

- DAVID MIRON – *Biology*

- JOEL MACAUSLAN – *Hard Sociology*

7:00PM–9:00PM EVENING SESSIONS
Session Chairs:

- HIROKI SAYAMA – *Concepts, Methods and Tools*

- RENE DOURSAT – *Neural and Physiological Dynamics*

- MARLENE WILLIAMSON – *Systems Engineering*

- *Socio-Economic Systems*

- COREY LOFDAHL – *Global Concerns*

WEDNESDAY, October 31

9:00AM–12:20PM MORNING SESSIONS
EVE MITLETON-KELLY – *Social Systems*

- KATHLEEN CARLEY – *Modeling Social Organization*

- MAX BOISOT – *Viewing China's Modernization Through the Complexity Lens*

- PAUL CILLIERS – *Complexity as Critical Philosophy*

- DWIGHT READ – *The Darwinian Route from Biological to Cultural Evolution*

2:00PM–5:00PM AFTERNOON SESSIONS
Session Chairs:

- DAVID MIRON – *Concepts, Methods and Tools*

- JONATHAN VOS POST – *Physical Systems*

- ALI MINAI – *Evolution*

- LEN TRONCALE – *Systems Engineering*

- IAN WILKINSON – *Management, Innovation and Marketing*

6:00PM–7:30PM BANQUET
7:30PM–9:00PM BANQUET SESSION
PAT HUGHES – *Global Challenges*

THURSDAY, November 1

9:00AM–12:20PM MORNING SESSIONS
DOUG NORMAN and SARAH SHEARD – *Social, Economic and Engineering Challenges*

- ROBERT DEAN – *Engineering Complex Systems*

- DON INGBER – *Principles of Bioinspired Engineering*

- RAY JACKENDOFF – *The Peculiar Logic of Value*

- RICHARD COOPER – *Global Economics*

2:00PM–5:00PM AFTERNOON SESSIONS
Session Chairs:

- GARY NELSON – *Scale Hierarchy*

- *Population Dynamics of Diseases*

- ANNE-MARIE GRISOGNO – *Systems Engineering*

- *Physical Systems*

- IRENE CONRAD – *Management, Innovation and Marketing*

7:00PM–9:00PM EVENING SESSIONS
Session Chairs:

- DAN BRAHA – *Networks*

- ALI MINAI – *Nueral and Physiological Dynamics*

- PHILIPPE BINDER – *Language*

- HELEN HARTE – *Social Policy for Science, Health and Education*

FRIDAY, November 2

7:30 AM–8:30AM HERBERT A. SIMON AWARD SESSION
MUHAMMAD YUNUS
9:00AM–5:00PM SPECIAL DAY ON NETWORKS
ALI MINAI – *Networks*

- ALAN PERELSON – *Networks of the Immune System*

- JOSE FERNANDO MENDES – *Evolution of Networks*

- JEFF STAMPS and JESSICA LIPNACK – *Networks and Virtual Teams*

- ALESSANDRO VESPIGNANI – *Epidemics on Networks*

- NICHOLAS CHRISTAKIS – *Social Networks and the Spread of Obesity*

- DAN BRAHA – *Dynamics of Networks*

Chapter 1

CALM: Complex Adaptive System (CAS)-Based Decision Support for Enabling Organizational Change

Richard M. Adler, PhD
DecisionPath, Inc.
rich@decpath.com

David J. Koehn, PhD
DJ Koehn Consulting Services, Inc.
koehndj@msn.com

Guiding organizations through transformational changes such as restructuring or adopting new technologies is a daunting task. Such changes generate workforce uncertainty, fear, and resistance, reducing morale, focus and performance. Conventional project management techniques fail to mitigate these disruptive effects, because social and individual changes are non-mechanistic, organic phenomena. CALM (for Change, Adaptation, Learning Model) is an innovative decision support system for enabling change based on CAS principles. CALM provides a low risk method for validating and refining change strategies that combines scenario planning techniques with "what-if" behavioral simulation. In essence, CALM "test drives" change strategies <u>before</u> rolling them out, allowing organizations to practice and learn from virtual rather than actual mistakes. This paper describes the CALM modeling methodology, including our metrics for measuring organizational readiness to respond to change and other major CALM scenario elements: prospective change strategies; alternate futures; and key situational dynamics. We then describe CALM's simulation engine for projecting scenario outcomes and its associated analytics. CALM's simulator unifies diverse behavioral simulation paradigms including: adaptive agents; system dynamics; Monte Carlo; event- and process-based techniques. CALM's embodiment of CAS dynamics helps organizations reduce risk and improve confidence and consistency in critical strategies for enabling transformations.

1 Introduction

Guiding organizations through transformational change is a daunting task. Examples of change include downsizings, mergers, and adopting new enterprise software systems or technology platforms. Transformational changes drive fundamental shifts in personal and organization ways of thinking and doing business. As such, they disrupt the status quo, forcing managers and workers out of their comfort zones, altering their mental models, and conflicting with established behavior patterns, processes, and cultural norms. Reactions resemble the body's immune response: organizations and individuals resist change and act to maintain prior "equilibrium" conditions, lowering morale, focus, and performance. Even if desired changes are implemented successfully, major challenges remain to *sustain* changes: absent ongoing vigilance, organizations tend to revert back to older, familiar behaviors and attitudes. In short, change, once effected, must be institutionalized to endure.

Businesses and government agencies report high failure rates in navigating, much less sustaining, transformations, even after spending considerable sums on change management consultants [1]. Diagnostic surveys reveal low levels of employee trust in management, and chronic dissatisfaction with working conditions. Unless these underlying root causes are addressed directly, management interventions to anticipate and mitigate the disruptive effects of transformational change will continue to fail.

This paper describes an innovative methodology for enabling and sustaining change called CALM™ (for Change, Adaptation, and Learning Model). CALM provides a low risk method for validating and refining change strategies. In essence, CALM "test drives" change strategies before rolling them out, allowing organizations to practice and learn from virtual rather than actual mistakes. Equally important, CALM allows organizations to monitor change strategies as they are being executed, and perform mid-course corrections as necessary. CALM enables change teams to sense and respond at the right moment, just as expert sailors tack quickly in response to small changes in the wind. CALM thereby reduces risk and improves confidence and consistency in transformational strategies.

2 Transformational Change

2.1 Related Work

The literature on change management is extensive, produced primarily by business school professors and management consultants [2]. Authors typically describe the disruptive forces and behaviors that they observe afflicting organizations facing major change, and then prescribe process-oriented methodologies designed to mitigate those problems. Kotter [3], for example, proposes the following process model. He argues that these eight phases are jointly necessary (but individually insufficient) to guide organizations through major changes and to sustain them.

- Initiate change (by defining a sense of urgency and business case for change)

- Build a coalition of change agents
- Formulate vision
- Communication and educate
- Empower others to act
- Create short-term wins
- Consolidate and further change
- Institutionalize change

Broadly speaking, change phenomenologies, while valuable for general anticipatory purposes, are purely qualitative and not directly actionable. We find additional problems with prescriptive process methodologies and how they are applied:

- Change management programs tend to focus at a tactical level, addressing a specific pending transformation rather than strategically, on *generalized* organizational preparedness and receptivity to continuous change
- Change process models apply the same mechanistic techniques that were developed to manage business and technology projects (e.g., scheduling, resource allocation) to the complex behaviors displayed by organizations and individuals facing disruptive changes — uncertainty, inertia, resistance
- Until recently [4], the literature effectively ignored decades of research on "new sciences" that focus expressly on modeling the kinds of social and personal behaviors observed in change situations, such as CAS, system dynamics, and stochastic methods. This latter literature, however, is not directly actionable.

2.1 CALM's Contribution: "Test Driving" Change Strategies

CALM addresses these problems by providing a *dynamic* decision support methodology. Rather than simply framing plans that encompass the eight phases of Kotter's change process, CALM introduces scenario-based situational modeling and "what-if" simulation capabilities to help organizations proactively validate and refine such plans. Specifically, CALM adds the following elements to the mix:

- A rich set of metrics designed to measure an organization's readiness to change
- A model of the dominant environmental forces, both internal and external, that influence organizations while they attempt to change
- A framework for defining organizational strategies – or *transformation plans* – comprised of change initiatives such as communication and compensation techniques that contribute to Kotter's archetypal phases for helping organizations and their workforces to accept and embrace significant changes
- A CAS-inspired dynamic model that estimates the (qualitative) effects of situational forces and change initiatives over time on CALM's readiness metrics

Consultants experienced with CALM facilitate meetings with teams of organizational leaders, managers and key employees to characterize the following:

- Organizational structure, key forces, and the pending (or ongoing) change
- Estimated current organizational readiness levels and specified target levels, which, if achieved would likely ensure success
- A small number of alternate plausible futures defined by assumptions of how environmental forces, trends, and singular events might play out
- One or more prospective alternate transformation plans, composed of pre-defined initiatives from CALM's library, which incorporate projected schedules and costs.

The CALM methodology is embodied in a software system implemented using DecisionPath's ForeTell framework [4].[1] ForeTell captures these various situational and assumptive elements in model constructs called *scenarios*. Consultants then apply CALM's simulation engine to project the likely outcomes of prospective plans — in terms of changing readiness metrics across alternate scenarios. Finally, integrated analytics enable projected outcomes to be explored and compared, across both alternate futures and transformation plans, to identify a robust change strategy.

Intuitively, a "robust" strategy is one that produces attractive results across alternate futures. No one can predict the future reliably. The next best thing is to devise a change plan that carries a high likelihood of success regardless of which future obtains. By projecting the likely consequences of candidate plans across diverse plausible futures, CALM helps organizations identify relative strengths and weaknesses, and uncover unintended consequences. Stronger plans can then be synthesized from preceding attempts. In short, the CALM framework supports the validation of iterative refinement of strategies to increase robustness.

Change strategies, however inspired, have little value unless they are executed competently. CALM was designed to be applied after strategies have been adopted, to monitor their execution. In this mode, organizations update their scenarios periodically to reflect current conditions and progress in improving readiness. CALM then re-projects the chosen strategy against these updated scenarios. If outcomes continue to be positive (i.e., readiness metrics reach their target levels), the chosen strategy is re-validated. If not, CALM acts as an Early Warning System, helping to uncover emerging problems promptly; diagnose them; and define and validate mid-course corrections. In this post-decision "sense and respond" mode, CALM helps organizations carry out and sustain change strategies across their extended "lifecycles," despite inevitable situational changes that occur over time.

2.2 Modeling the Behavioral Dynamics of Organizational Change

Enabling and sustaining transformational change is a complex, extended process that is influenced by several basic dynamic drivers. First, as organizations carry out change initiatives, internal and external stakeholders invariably respond, adapting

---

[1] ForeTell is a software platform for rapidly developing and deploying critical decision support systems. It has been applied to diverse domains including counter-terrorism and pandemic preparedness, competitive drug marketing strategy, IT portfolio management, and dynamic social network analysis.

their behaviors to advance personal and group interests.

Second, individual and organizational attitudes and behaviors tend to evolve over time in complex, and often non-linear patterns. For example, trust, morale, focus, and acceptance typically don't build or decay continuously and smoothly; rather, they tend to jerk, stick, and accelerate or decelerate.

Finally, environments such as societies and markets evolve continually, driven by situational forces and events. In short, the target audience and the "ground" under the organization's "feet" shift continually and in ways that are difficult to anticipate. It is the complex interplay of intentional adaptive and non-linear behaviors and ongoing environmental change that causes static process models to fail.

CALM improves how transformational plans are developed, validated, and executed because it recognizes and embraces these complex dynamics of change. It applies "new science" theories such as system dynamics, CAS, and Monte Carlo methods specifically designed to model individual and social behavior patterns. Finally, it provides an iterative and interactive process for modeling and analyzing change and change strategies using these "organic" dynamic modeling techniques.

A simple analogy explains CALM's approach and differentiate it from process models such as Kotter's. Enabling and sustaining transformational change is similar to launching a rocket. Propelling a rocket into orbit requires generating sufficient over time to (1) overcome inertia and lift the rocket's mass, and (2) accelerate it to "escape" velocity for insertion into orbit despite the constant drag of gravity. If insufficient thrust is generated and maintained, the rocket will fall back to earth. In multi-stage rockets, the ignition and firing durations of booster sections must be designed precisely to produce sufficient thrust and appropriate acceleration profiles.

Similarly, change strategies must generate organizational and personal "thrust" to bring out (i.e. accept and embrace) change *and* sustain it by overcoming persistent conservative forces such as personal and group inertia and old behavior patterns. As in multi-stage rocket engines, the sequencing and durations of change activities must be carefully coordinated to prevent the loss of momentum and perceptions of stalled progress, which lead directly to outright failure or unraveling of transformations.

This analogy is admittedly imperfect. Designing rocket engines and launch trajectories is a well-established engineering discipline. Newton's laws of motion strictly determine the mechanical interactions of a small number of key parameters and forces. Solutions can be computed from textbooks or software programs. The same equations apply uniformly to all launch situations. They never change over time.

Organizational change clearly represents a more complex and open-ended phenomenon. It is not obvious what parameters to measure, much less what, if any "universal laws" govern situational dynamics. As a result, models such as CALM are qualitative rather than quantitative, and *exploratory* than deterministic and predictive.

Process models such as Kotter's, while valuable, provide guidance that is largely static, passive, and broadly defined: they offer no framework for thinking through the *dynamics* of change processes, or help in designing or selecting and assembling specific initiatives to implement the eight stages tailored to particular organizations and their specific transformational challenges.

Most organizations need more detailed guidance. Enabling change is a complex undertaking. Mistakes are inevitable. They may also be irreversible. Initiatives that fail tend to undermine stakeholder trust and confidence: management cannot simply switch strategies and try again from the same initial state. Something must be added to help organizations design and test change strategies in advance.

Our rocket analogy provides this critical missing ingredient, namely, a model for anticipating how transformational change is likely to play out in terms of empirical metrics. CALM equates Kotter's eight step model to a multi-stage rocket engine. CALM also defines an explicit dynamic model of how change initiatives (and situational forces) impact measurable organizational readiness factors. This simulation model allows CALM to project how organizations, their employees, forces, and change initiatives will interact with one another and evolve over time.

Transformation plans can thereby be validated dynamically, much as rockets can be simulated to see if they generate the required launch and thrust profiles. For CALM, sufficient "thrust" over time equates to improvements in the key readiness metrics. Reaching escape velocity corresponds to achieving target readiness values. Failure to achieve target values means that the transformation plan is unlikely to succeed.

CALM's simulation, while qualitative, provides a systematic and repeatable basis for projecting likely outcomes. Through iterative refinement (and continuous improvement over successive transformations), CALM helps organizations address change and learn how to execute better over time. The next section describes the software that embodies this methodology.

3 Embodying the CALM Methodology in Software

CALM organizes the process of "test driving" change strategies into three primary processes – modeling, behavioral "what-if" simulation, and analysis. Each process is supported by a software application that was developed using DecisionPath's ForeTell® decision support software platform [5], as summarized in Figure 1.

7

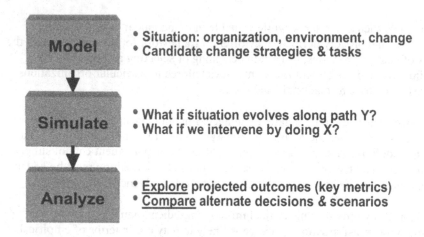

Model	• Situation: organization, environment, change • Candidate change strategies & tasks
Simulate	• What if situation evolves along path Y? • What if we intervene by doing X?
Analyze	• <u>Explore</u> projected outcomes (key metrics) • <u>Compare</u> alternate decisions & scenarios

Figure 1. Phases of CALM Methodology

3.1 Modeling

CALM explicitly captures available knowledge about both static *and* dynamic aspects of situational contexts and transformations. Capturing situational dynamics is clearly critical for CALM's second task – projecting outcomes.[2]

CALM change models, called *scenarios,* are comprehensive, encompassing the following situational elements mentioned above,[3] summarized pictorially in Figure 2:

- Relevant organizational units (e.g. an agency, its member departments or groups)
- The transformation at issue (e.g., new IT system technology, a reorganization)
- Current and target (goal) values for metrics that measure organizational readiness to change (and overall performance)
- Environmental forces and trends acting on the organizational unit on a continuing basis (and assumptions about how they are likely to change over the future)
- Possible events that might occur and disrupt the transformation and change plan
- Candidate transformation plans, composed of individual change initiatives

Environmental forces include both external factors (e.g. social, political, legal,

---

[2] Lacking supporting software, conventional change management methodologies address dynamics in an informal, intuitive manner, if at all. This ad hoc approach is very difficult to apply consistently, much less replicate or teach.

[3] These elements are implemented as a hierarchical object-oriented model: the Scenario class contains (is the parent of) Organizational Unit, which contains Change, Metric, Environmental Force, and Transformation Plan subclasses.

economic) and internal factors (e.g., leadership, resources). Examples of disruptive events include new legislation or regulations, changes in leadership or economic conditions.

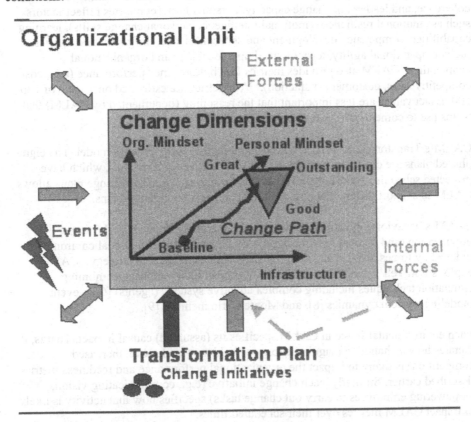

Figure 2. CALM Elements for Modeling Organizational Change

Metrics are critical for measuring an organization's initial, goal, and evolving states surrounding a transformation. Existing change methodologies focus primarily on metrics relating to business "infrastructure", which measure "physical things to do" regarding organizational technology processes, structures, and policies (e.g., functional training to use a new business software system). However, the dominant causes for transformation failures typically trace back to less obvious and less concrete organizational and workforce readiness factors, such as cultural dissonance; inadequate leadership support; poor communication; fear of change; lack of confidence; or inadequate skills or empowerment to carry out the necessary adaptations. Failure to attend to these intangible factors can result in winning the battle, but losing the war: organizations know what to do to carry out the requisite changes, but are unwilling to accept or embrace them.

The CALM methodology measures organizational readiness to deal effectively with change across three "dimensions" – infrastructure, organizational and personal (workforce) mindsets. Each CALM change readiness dimension consists of three to six metrics, selected from the literature to reflect diverse types of transformations and associated risks. Example organizational mindset metrics include teaming, cultural coherence, and leadership change capacity. Personal mindset metrics reflect factors such as emotional resilience (emotional intelligence/personal change skills), learning capabilities, competency development and self-confidence. Infrastructure metrics include operational agility, technological upgradeability and organizational adaptability. CALM also provides metrics for "bottom-line" performance (e.g., cost competitiveness, customer satisfaction).[4] All metrics are estimated on a scale of 1 to 100. Exact values are less important that the reasoning (documented by CALM) that teams use to come to consensus on them.

CALM's Transformation Plan construct exploits the Kotter process model. The eight-phased plans are constructed of finer-grained initiatives or strategies, which have projected schedules and estimated costs. Incorporating costs into change plans allows CALM to support differential cost-benefit analyses of competing plans.

CALM's behavioral dynamics model projects the likely impacts on the target organization (and its readiness measures), of changes in the situational environment induced by forces, events, and the proposed change enablement strategy. CALM exploits ForeTell's dynamic architecture to project likely impacts, combining simulation techniques including complex adaptive systems (agents) [6,7], event modeling, system dynamics [8], and Monte Carlo methods [9].

Each environmental Force in CALM specifies its (assumed) causal impact. That is, it dictates how a change in magnitude (e.g., of bold new leadership, increased competition) is likely to impact the organizational performance and readiness metrics described earlier. Similarly, each change initiative (e.g., communicating vision, empowering employees to carry out change tasks) specifies how that activity is likely to impact CALM metrics over their scheduled times.

As in the real world, these effects may take time to manifest (i.e. delay or latency), have bounded durations, and display cumulative build-up (i.e., non-linear response). The data to drive these specifications comes from expert judgment and other sources (e.g., surveys, benchmarking exercises, data warehouses) that may be available. Users construct, copy and edit, and export scenarios using an intuitive graphical user interface (GUI) modeled after standard tree-based controls for viewing file systems. The Scenario Editor contains panels that depicts the entities comprising a scenario, the attributes for a selected entity, and integrated help that explains the entity type and attribute. Pop-up editors provide text, numerical slider, list and table controls for entering scenario data, as appropriate. All values can be annotated with comments, source, and degree of certainty, facilitating transfer and maintenance of scenarios.

For ease-of-use and speed, we are populating a library of pre-defined forces and

---

[4] The CALM software allows the baseline set to be extended with additional metrics tailored to particular organizations, sectors, and changes.

change initiatives integrating with CALM. Each such component provides a pre-validated causal model, which can be tuned to specific organizations and contexts as required. CALM's library allows drives a "Lego™ building block" approach to building scenarios that exploits best practices expert knowledge about change strategies and their likely impacts on organizational readiness over time.

3.2 Simulation Engine

CALM projects situational dynamics via ForeTell's parallel discrete event simulation engine. The core engine employs an agent-based simulation paradigm: at each simulated interval (i.e. "clock tick"), this engine invokes active scenario entities in a uniform order based on the object hierarchy. Each such entity runs its type-specific behaviors. For intentional entities (i.e., goal-directed organizations or persons), actions often involve sensing internal and external state and responding according to behavioral patterns, such as stimulus-response rules.

The ForeTell engine employs a novel hybrid architecture that extends the core CAS paradigm with overlays that support other simulated dynamics, including causality (i.e., system dynamics), situational trends, events, and processes. For example, a CALM transformation plan carries out organizational change initiatives as a process comprised of scheduled tasks, the causal impact of which is realized by system dynamic productions.[5] A similar embedded system dynamics model propagates the effects of CALM environmental forces on organizational readiness metrics.

ForeTell's multi-modal simulator reflects the fact that real world systems are influenced by diverse — and interacting — dynamic drivers. Uni-modal simulators frequently lead to distorted or incomplete models when their chosen dynamic paradigm is applied outside of its ideal design stance (e.g., modeling environmental interactions with agents or systems of intentional entities with system dynamics).

Users monitor and control executing scenarios through ForeTell's "dashboard" style GUI, made up of controls, gauges and time series plots of aggregate readiness metrics. Simulations can be suspended to inspect specific entities and metrics.

3.3 Analytics

As the ForeTell simulator runs, it logs all state changes occurring in CALM scenario entities to a database. An integrated analytics engine helps users retrieve and reduce this mass of data via graphical and tabular summaries to explore projected outcomes of individual CALM scenarios. More importantly, outcomes can be compared across scenarios for competing strategies and/or diverse plausible futures.

Users generate summary analytics via a simple menu-driven dynamic query GUI.

---

[5] ForeTell's simulator incorporates a Monte Carlo utility that allows users to define statistical distributes of scenario parameters and perform large numbers of runs (trials) in batch mode, which is useful for sensitivity analysis. It also supports dynamic social network modeling and Bayesian inference networks.

Outputs include tabular reports, time series and radar plots, and frequency histograms. One summary report documents before/after metrics and percentage change, while others isolate specific dynamics (events, trends, causality) that help users understand why and when observed readiness changes occurred. ForeTell's analytics engine embeds open source math, graphics, and statistics libraries, allowing rapid extension to satisfy new analytic requirements.

3.4 Validation and Verification

Organizations typically require confidence building exercises before they are willing to commit to the CALM methodology. CALM employs several techniques for this purpose. First, when available, we use an organization's prior transformational projects to calibrate CALM's dynamics, by adjusting force and change initiative causal weighting factors to match historical patterns of changing readiness levels.

Secondly, the CALM methodology prescribes two types of validation exercises, called "sanity checking" and retrospective testing. In sanity checking, scenarios are constructed to depict extreme conditions and change strategies. People often lack firm intuitions about the minor variations in forces or plans. However, they typically have stronger instincts about outcomes in extreme situations.

For example, suppose employees face a change situation that entails deep pay cuts and/or layoffs. Given this context, most people would predict with confidence an outcome that includes major drops in employee morale and customer satisfaction. An organization's CALM scenarios should project outcomes that match these beliefs. They should also change gracefully as the extreme conditions are "dialed back."

Retrospective testing on past change situations represents the "acid test". Here, scenarios are created that depict the organization's state and environment at the (historical) point of decision. Next, the forces, trends and events that actually occurred through to the present are added to the scenario. Finally, the change initiatives that were undertaken are introduced. Given these inputs, CALM's scenario projections should resemble the evolution of organizational readiness metrics that actually took place.

4.0 Conclusions

CALM provides a disciplined framework drawn from CAS principles for modeling organizational transformations and exploring alternate change enablement strategies. CALM acts as a "robotic juggler", uniformly manipulating the diverse interacting dynamic drivers of change that humans are cognitively unable to project mentally in a consistent manner. In essence, CALM enables organizations to practice prospective change strategies in a low risk virtual environment. Organizations can then learn from simulated mistakes at minimal cost, rather from real "blood on the tracks" errors that result in problems such as pervasive worker mistrust or alienated customers.

CALM can be applied over the lifecycle of extended transformations, helping organizations sense and respond to continuous situational change. CALM also supports decision-making for less radical incremental or transitional forms of organizational change.

CALM's differential analyses of alternate change strategies across diverse plausible futures are key drivers of enhanced decision-making. All significant changes involve risk: risk is unavoidable. Managing risk <u>effectively</u> hinges on understanding the likely costs and benefits of assuming particular risks and incurring only those risks for which the rewards are commensurate. CALM helps organizations explore these trade-offs systematically and repeatably. CALM libraries capture and disseminate best practice analyses of change and change strategies. CALM simulations also provide audit trail that enhance organizational governance, continuous learning and improvement. Thus, the CALM methodology helps organizations reduce risk and increase confidence in responding to transformational change.

Bibliography

[1] Pascale, R., Millemann, M., and Gioja, L. Surfing the Edge of Chaos, p. 12.

[2] Useful Web-based bibliographies for change management literature include: http://www.dhrm.state.va.us/training/change/resources_bibliography.html, http://www.change-management.com/bookstore.htm, http://www2.nrcan.gc.ca/es/msd/emmic/web-en-emmic/index.cfm?fuseaction=subjects.subjectchange.

[3] Kotter, J.P. Leading Change: Eight Ways Organizational Transformations Fail. Harvard Business Review, March-April, 1995, 59-67.

[4] Herasymowych, M. and Senko, H., Navigating through Complexity: Systems Thinking Guide, 2nd Edition, 2002, www.mhainstitute.ca.

[5] Adler, R. M., "ForeTell: A Simulation-Based Modeling and Analysis Platform for Homeland Security Decision Support". Proceedings Second IEEE Conf. on Technologies for Homeland Security. Cambridge, MA. May, 2003.

[6] Bar-Yam, Y. Dynamics of Complex Systems, Westview Press, Cambridge, MA 1997.

[7] Epstein, J.M. and Axtell, R., Growing Artificial Societies: Social Science from the Bottom Up. Brookings Institution Press, Washington DC, 1996.

[8] Sterman, J., Business Dynamics: Systems Thinking and Modeling for a Complex World, Cambridge, MA, Irwin McGraw-Hill, 2000.

[9] Fishman, G.S., Monte Carlo: Concepts, Algorithms, and Applications, New York, Springer, 1996.

Chapter 2

Multi-level behaviours in agent-based simulation: colonic crypt cell populations

Chih-Chun Chen
Department of Computer Science,
University College London,
c.chen@cs.ucl.ac.uk
Sylvia B. Nagl
Department of Oncology and Biochemistry,
University College London,
s.nagl@medsch.ucl.ac.uk
Christopher D. Clack
Department of Computer Science,
University College London,
c.clack@cs.ucl.ac.uk

Agent-based modelling and simulation is now beginning to establish itself as a suitable technique for studying biological systems. However, a major issue in using agent-based simulations to study complex systems such as those in Systems Biology is the fact that simulations are 'opaque'. While we have knowledge of individuals' behaviour through agent rules and have techniques for evaluating global behaviour by aggregating the states of individuals, methods for identifying the interactive mechanisms giving rise to this global behaviour are lacking. Formulating precise hypotheses about these multi-level behaviours is also difficult without an established formalism for describing them. The complex event formalism allows relationships between agent-rule-generated events to be defined so that behaviours at different levels of abstraction to be described. Complex event types define categories of these behaviours, which can then be detected in simulation, giving us computational method for distinguishing between alternative interactive mechanisms underlying a higher level behaviour. We apply the complex event formalism to an agent-based model of cell populations in the colonic crypt and demonstrate how competition and selection events can be identified in simulation at both the individual and clonal level, allowing us to computationally test hypotheses about the interactive mechanisms underlying a clone's success.

1 Introduction

Biological systems are complex adaptive systems (CAS) where a great number of entities interact to give rise to system-level behaviours and processes. These systems are inherently difficult to study because they exhibit polymorphism, context dependency, evolution, reprogrammability, emergence, non-linearity, heterogeneity, hierarchy and complexity [7], [13], characteristics shared by most complex systems. Agent-based modelling and simulation (ABMS) is now a fairly well-established technique for studying such complex systems [16] but it has only recently begun to be seriously adopted in Systems Biology e.g. [17]. When used to study biological systems, ABMS allows certain hypotheses about individual-level behaviour (e.g. at the level of cells) to be validated and refined, since the overall system behaviour observed in simulation can then be compared with that observed in the real system. While a correspondence does not verify a hypothesis, it does show that it is valid and able to generate the expected behaviour. ABMS is therefore seen as a way of performing 'thought experiments' [12]. Rules are defined at the agent level, while the behaviour of the whole system is typically represented by a macro-state variable that aggregates the states of all the agents in some way; this macro-state variable is then tracked through time.

A major problem with this approach is the loss of structure when states are aggregated e.g. no information about spatial locality is retained. This means that we are unable to identify the mechanisms (the actual interactive patterns between agents) that give rise to a particular global behaviour. For this reason, simulations are usually visualised, allowing the human experimenter to observe visually the interactions taking place through time. Hypotheses about such interactions are then formulated in natural language and hence vague e.g. 'the cells cooperate to survive'. In this paper, we seek to address this problem by introducing a formalism that allows such hypotheses to be expressed precisely in terms of the agent model. Once expressed formally, we can then identify the particular interactive mechanisms or classes of mechanisms in an agent-based simulation, giving us a computational method for testing such hypotheses. We illustrate this using ABMS of cell populations in the colonic crypt.

The section that follows (Section 2) briefly introduces the complex event formalism. Section 3 describes the agent-based model of colonic crypt cell populations. Section 4 formulates hypotheses about clonal level behaviours using the complex event formalism and discusses the results from detecting these behaviours in simulations. The final section (Section 5) concludes the summarises and concludes the paper.

2 Compositionality and Complex events

In this section, we briefly introduce the complex event formalism, which allows multi-level behaviours in agent-based simulations to be described. These behaviours are sometimes called 'emergent' because they have organisational properties that are not explicitly specified in the agent rules (reviews of theories of emergence can be found in [6], [5] and [1]). A more detailed account of the formalism and its relevance to current theories of emergence can be found in [3].

There are four central ideas behind the formalism, all relating to way that properties (in this case behaviours) can be located in a system or simulation.

- Every behaviour in a system can be described by events (state changes) located in an n-dimensional (hyper)space. For the lifetime of the system or simulation, events can be located in this space by specifying the coordinates in each of the dimensions. The coordinate system used to specify the location can be global (from a whole system point of view) or local (where locations are in relation to a particular constituent *within* the system). [1]

- If two macro-properties consist of constituents of the same types and constituents of the same type have the same configuration with respect to each other in the two properties, we can say the two properties are of the same type.

- We can describe **regions** as well as point locations in a system or subsystem space using propositional statements about the location in the system/simulation's various dimensions. For example, in a system with only time and identity represented, (before 3, 4) stands for all the states or state transitions that occur in component 4 before time step 3.

- Higher level properties can be composed by defining organisational relationships between their constituents i.e. their **configuration**. This idea is generalisable to any dimension.

2.1 Complex events and simple events

In an agent-based simulation, every event is the result of an agent rule being applied; we call these simple events. Simple events can be defined at various levels of abstraction, depending on which of the components (e.g. variables, agents[2]) affected by the rule application we are concerned with. For example, a

---

[1] For example, if the global coordinate (12, 1, 4, 2) represents the location of a state transition in the 12th time step (first tuple item holds time), located in coordinate (1, 4) of physical space (second and third tuple items hold space) in component with ID 2 (final tuple item holds component identity); the equivalent coordinate using a local coordinate system defined with respect to component 2 at time step 11 in the same spatial location would be (1, 0, 0, 0).

[2] Agents can be treated as complex variables.

16

rule that causes state changes in components a, b and c can cause simple events $(q_a, q_b, q_c) \rightarrow (q'_a, q'_b, q'_c)$, $(q_a, q_b) \rightarrow (q'_a, q'_b)$..., $q_a \rightarrow q'_a$...etc. We call this the scope of the event. Two simple events e_1 and e_2 in a system are said to be of the same type if (a) e_1 and e_2 result from the same agent rule and (b) the scope of e_1 is identical to the scope of e_2 i.e. for every component in which a state change occurs in e_1, there is a component of the same type in which the same type of state change occurs in e_2[3].

A complex event CE is defined as either a simple event SE or two complex events linked by $\bowtie$:

$$CE \ :: \ SE \ | \ CE_1 \bowtie CE_2 \tag{1}$$

$\bowtie$ denotes the fact that CE_2 satisfies a set of location constraints with respect to CE_1. Conceptually, complex events are a configuration of simple events where each component event can be located in a region or point in a hyperspace that includes time, physical space and any other dimensions. The set of location constraints can be represented as a coloured multi-graph, where the node colours stand for event types and the edge colours for different relationship types (the location constraints) existing between the events [4].

2.2 Complex event types for multi-level behaviour

We have already introduced the idea that events can be typed in our discussion of simple events. We now extend this to complex events. Two complex events CE_1 and CE_2 are said to be of the same type if, for each constituent event $e1$ in CE_1 there is exactly one event $e2$ in CE_2 satisfying the same location constraints, and $e1$ and $e2$ are events of the same type. To specify a complex event type therefore, we need to specify the types for each of the constituent events and the location constraints that hold between them.

Complex event types can differ in specificity. A fully determined complex event type CET_{Full} is defined as one whose constituent events are in a fully determined configuration i.e. given the global location of one constituent event in the complex event, it is possible to work out the precise location of every other constituent event. A partially determined complex event type CET_{Part} is an event type with a partially determined configuration and therefore defines a set of complex events with fully determined configurations.

$$CET_{Part} = \{CET_{Full}\} \tag{2}$$

The dimensions in which configurations are not fully specified lower the resolution of the complex event, with weaker constraints (greater ranges of possible values) implying a lower resolution in that dimension. More generally, the greater the number of complex event types with fully determined configurations that a complex event type contains, the lower its resolution.

---

[3] See [9] for a formal definition of types

Having briefly outlined the complex event formalism, we now introduce the agent-based model of colonic crypt cell populations used to demonstrate its application.

3 Cell population model

In this section, we describe the agent-based model used for simulations of colonic crypt cell populations. Section 3.1 gives the biological background that forms the basis of the model while Section 3.2 gives the agent rules.

3.1 Biological background

The colon is made up of villi, which are finger-like structures each made up of ~300 cells - 15 cells in diameter, 20 cells from the closed bottom (colonic crypt) to the villus tip [14]. In a colonic crypt, cells divide, differentiate and migrate up the crypt. Stem cells reside at the bottom of the crypt and typically divide asymmetrically to give one transit cell and one stem cell. Transit cells have the ability to divide a limited number of times (usually around 3 times) after which they undergo terminal differentiation. Fully differentiated cells are removed from the luminal surface by programmed cell death (apoptosis).

Cells can take two to seven days to migrate from the site of their final division to the villus tip [18] and stem cells have cycle times ranging from 10 to 14 hours (consisting of G_1, S, DNA repair, G_2, and M phases) after which they enter a resting phase (G_0) of one one or two days before they divide again [2].

3.2 Rules for cell agents

In our simulations, a single time step represents an hour of real time. Given what we know about the durations of cell division and migratory processes (as outlined in Section 3.1), Table 1 summarises the event timings used in the model. Where a range is given, a duration within the range is randomly selected.

Table 1: Durations for cell cycle stages and migration

Event	Real time duration (range)	Simulation time steps
G_0 (Resting)	24h-48h	24-48
G_1 Phase	1h-5h	1-5
S Phase	8h	8
Repairing DNA	1h-5h	1-5
G_2 Phase and M Phase	1h	1
Migration	48h-168h	48-168

18

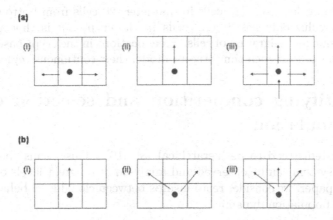

Figure 1: (a) Possible target locations for insertion when a new cell agent is produced from cell division. The arrows represent the possible locations for insertion if the parent cell is located in the position occupied by the black dot. (i) The new cell first attempts to insert itself in each of the adjacent positions (the order is determined randomly). (ii) If the two adjacent locations are both occupied, the new cell tries the location directly above the parent cell. (iii) If all these positions are occupied, it randomly selects one of the occupied positions (including the parent cells) and attempts to oust the cell currently occupying that position. The parent cell itself might be ousted if its position is randomly selected by the newly produced cell. If the cell fails to oust the existing cell, it fails to be inserted and 'dies. **(b)** Migration. (i) The cell agent first tries to move into the position directly above its current location. (ii) If the position directly above is occupied, it tries each of the positions adjacent to this (the order is determined randomly). (iii) If all these locations are occupied, the cell randomly selects one of the occupied positions and tries to oust the cell currently occupying that position. If it fails, the cell remains where it is.

When a cell agent divides, it produces an additional cell agent which needs to be inserted at a location close to the parent cell. If all the locations in the parent cell's neighbourhood are occupied, the new cell agent randomly selects a location that it attempts to occupy by 'killing' the cell currently occupying the location (this might be the parent cell). Similarly, when a cell attempts to migrate upwards, it can 'kill' a cell occupying the space it is trying to move into (see Figure 1). The likelihood of a cell agent ousting another is a function of its fitness *relative* to its competitor's so that two cells with equal fitnesses have the same probability of 'losing' in a competition (in the simulations presented here, all cells have equal fitnesses). After three divisions, transit cells differentiate and can no longer divide, although they continue to migrate before eventually being lost when they reach the villus tip.

The model is used to simulate a single villus with maximum capacity 300 cells i.e. 300 grid locations (15 cells in diameter, 20 cells from bottom to top). Initially, the villus only has 6 stem cells in the crypt but is then populated through division and migration of cells. Free locations in the crypt base are filled with new stem cells at each time step to model their continuous replenishment.

4 Identifying competition and selection events in simulation

Biological systems tend to be hierarchical [8], [10]. This means that the behaviour of a system can be observed and analysed at different levels of abstraction. In this paper, we consider relationships between clonal level behaviour and overall clone population dynamics.

4.1 Clone population dynamics

Ten 1000-time-step simulations were run based on the model. For each simulation, the clonal populations were tracked through time, with several types of dynamics being observed (see Appendix 1). In some cases, one or two clones were significantly more successful than the others. (This result in itself has interesting implications for the Systems Biology of Cancer, since it means that tumours may develop even when their cells have no intrinsic selective advantage over normal cells, supporting the hypothesis that cancer is often a systems disease [11]).

We try to validate two hypotheses about the strategies clones adopt to when they are expanding. We consider the significance of clustering and a particular 'motif' of behaviour that it enables. The section that follows demonstrates how complex event types can be used to identify these.

4.2 Complex event types for clustering structure and clonal behaviour

We wish to test two hypotheses:

1. Clustering is common in successful clones (or when clones are successful[4]) since members of the clone protect one another.

2. Move events by members of a clone can lead to clonal expansion when a division or move by another clone member means the unoccupied previous location is filled by another clone member (see Figure 2). We hypothesise that this pattern of behaviour is more common in successful clones/when clones are successful.

These can be reformulated in terms of complex events so that they can be tested computationally during simulation. We begin by defining the following simple event types, each with a particular scope:

---

[4]As mentioned above, clone success often varies throughout the simulation.

Figure 2: move-win-replace complex event type. First, a cell belonging to a clone cluster moves into a new location and ousts a cell belonging to another clone. Then the cell's previous location is occupied by another cell in the clone cluster, allowing the cluster to expand.

- ma: cell attempts to move into a new location. The scope consists of (i) S, the source location; (ii) T, the target location; and (iii) M, the moving agent.

- ia: cell attempts to occupy a location currently occupied by another cell. The scope consists of (i) I, the cell attempting to occupy the location; and (ii) O, the cell currently occupying the location.

- cp: competition between two cells. The scope consists of (i) W, the cell that wins in the competition; and (ii) L, the cell that loses in the competition.

- mv: cell moves to a new location. The scope consists of (i) S, the source location; (ii) T, the target location; and (iii) M, the moving agent.

- dv: cell divides to give a new cell. The scope consists of (i) P, the parent cell and (ii) C, the child.

- in: a newly created cell is inserted into a location. The scope consists of (i) N, the new cell and (ii) T, the target location.

Within-clone competition   Clone clustering can be determined by the degree of within-clone competition, which corresponds to the event type:

$$wcc :: cp(W.cloneID = L.cloneID) \tag{3}$$

This is a sub-type of the *compete* event type.

Move-win-replace   A strategy that is believed to be successful for expanding a clone cluster is where a cells first replaces a cell from another clone and then has its own previous location filled by a member of its own clone. This corresponds to the complex event type (see Figure 2):

$$mwr :: ma1 \bowtie_A ia \bowtie_B cp \bowtie_C mv1(\bowtie_D ma2 \bowtie_E mv2 \| \bowtie_F dv1 \bowtie_G in)$$

where

- $\bowtie_A::$ $(; [T_{ma1} = O.loc, M_{ma1} = I])$, i.e. the target location T_{ma1} of the move attempt $ma1$ is the same as the occupant O's location in the invade attempt ia and the moving agent M_{ma1} is the invader I.

- $\bowtie_B::$ $(; [I = W])$, i.e. the invader I in invade attempt ia is the winner W in the compete event cp.

- $\bowtie_C$:: $(; [L.loc = T_{mv1}, W = M_{mv1}])$, i.e. the loser L's location is the target location T_{mv1} of the move event $m1$ and the winner W in the compete event cp is the moving agent M_{m1} in the move event $mv1$.move is winner.

- $\bowtie_D$:: $(; [S_{mv1} = T_{ma2}, M_{mv1}.cloneID = M_{ma2.cloneID}, M_{m1}! = M_{ma2}])$, i.e. the target location T_{m1} of the second move attempt $ma2$ is the same as the source location S_{m1} of the first move event $m1$, the moving agents M_{m1} and M_{ma2} belong to the same clone but are different indivduals.

- $\bowtie_E$:: $(; [T_{ma2} = T_{mv2}, M_{ma2} = M_{mv2}])$, i.e. the target T_{ma2} of the move attempt $ma2$ is the same as the target T_{mv2} of the actual move $mv2$ and the moving agent is the same individual.

- $\bowtie_F$:: $(; [M_{mv1}.cloneID = C_{dv}, M_{mv1}! = C_{dv}])$, i.e. the moving agent M_{mv1} and the newly created child agent C_{dv} belong to the same clone but are different individuals.

- $\bowtie_G$:: $(; [C_{dv} = N_{in}, S_{mv1} = T_{in}])$, i.e. the target T_{in} of the insert is the same as the source location S_{mv1} of the move $mv1$ and the new cell from division is the same cell as the new cell to be inserted.

(; is the next event operator.[5])

The complex event mw stands for the complex event where a cell moves into a space previously occupied by a cell from another clone:

$$mw :: ma1 \bowtie_A ia \bowtie_B cp \bowtie_C mv$$

We can now re-formulate our two hypotheses in complex event terms:

1. When a clone is successful (has a greater number of individuals compared to other clones), it will have (proportionally) more within-clone competition wcc events (relative to the overall number of competition events for the clone), indicating that more of its cells exist in a cluster.

2. A successful clone will have (proportionally) more move-win-replace mwr events (relative to the number of mw events for the clone).

Since it is highly contested whether the classical model of causality holds for complex systems, we do not make reference to it in our hypotheses, nor do we assume it.

4.3 Results and discussion

To validate the two hypotheses, we first considered clonal success at 100 time-step intervals. Clonal success is represented by the average number of individuals in each clone μ_X over a time interval ρ divided by the overall average number of individuals μ_{ALL} i.e. for each clone X:

---

[5]In this particular example, we are assuming the simulation is a discrete event simulation so that each step does not necessarily have to represent the same unit of time.

$$success_\rho = \frac{\mu_X}{\mu_{ALL}}, \qquad (4)$$

where

$$\mu_X = \frac{(v_X t_m + v_X t_{m+1} \ldots + v_X t_n)}{n - m},$$

$$\mu_{ALL} = \frac{\sum(\mu_{X1}, \ldots \mu_{X\sigma})}{\sigma},$$

$\rho = n - m$ is time interval (in the anlysis presented here, $\rho = 100$), $v_X t_i$ is the number of individuals in X at time step i), and σ is the number of clones. $success_\rho$ therefore indicates each clone's success relative to the others in a given simulation. The proportions of wcc events (relative to $success_\rho$) and mwr events (relative to $success_\rho$) were then calculated for these intervals. This was done for each simulation and then for the whole set of simulations. Results of clones that became extinct were omitted after the time interval in which they became extinct.

The respective relationships between clonal success and wcc/mwr were evaluated by calculating the correlation coefficient r between clonal success and wcc/mwr occurrence (see hypotheses above). A t-test was then conducted for each of these to test their significance. For the single simulation analyses,

Since each simulation had 1000 time steps and 6 clones, the total number of data items considered was $N \leq 60$ for the single simulation analyses (less if there were extinctions) and 1200 for the aggregated analysis. The results are shown in Table 2.

Table 2: Correlations between clonal success and wcc events/mwr events. Accuracy 3 decimal places. The p values do not assume directionality and significance (sig.) is determined based on a 95% confidence interval.

Sim.	N	df	r_{wcc}	t_{wcc}	p_{wcc}	r_{mwr}	t_{mwr}	p_{mwr}
1	60	58	0.277	2.196	0.032 (sig.)	-0.386	-3.183	0.002 (sig.)
2	44	42	0.301	2.043	0.047 (sig.)	-0.402	-2.845	0.007 (sig.)
3	37	35	0.212	1.285	0.207	-0.286	-1.767	0.086
4	46	44	0.254	1.739	0.089	-0.378	-2.712	0.010
5	52	50	0.242	1.763	0.084	-0.474	-3.807	0.000 (sig.)
6	60	58	0.298	2.378	0.021 (sig.)	-0.307	-2.458	0.017 (sig.)
7	43	41	0.270	1.796	0.080	-0.477	-3.479	0.001 (sig.)
8	51	49	0.333	2.469	0.017 (sig.)	-0.422	-3.257	0.002 (sig.)
9	47	45	-0.110	-0.742	0.462	-0.237	-1.634	0.109
10	53	51	0.234	1.719	0.092	-0.300	-2.247	0.027 (sig.)
All	493	491	0.152	3.420	0.001 (sig.)	-0.179	-4.025	0.000 (sig.)

Overall and in four out of the ten simulations, the (positive) correlation r_{wcc} between wcc events and clonal success was significant. There was also a significant correlation between mwr events and clonal success overall and in six of the simulations, but the direction was negative, the opposite to that hypothesised. This latter result is counter-intuitive and will be investigated in a future paper since it requires analysis using other complex event types. As well as considering different simulations, we also carried out a correlation analysis for each clone in each simulation (these results are given in Appendix 2). Again, wcc events and mwr events correlated with clonal success on some occasions but not others. The differences in the t values for r_{wcc} and r_{mwr} (determining their significance) for the different simulations and clones implies that the same global effect (e.g. clonal success) can have different underlying mechanisms, even with the *same* agent-based model. The next step would be to determine which other mechanisms are at work and which mechanisms tend to correlate with one another. This can again be done through the specification and detection of complex event types.

5 Summary and conclusions

In this paper we have shown how the complex event formalism can be used to specify multi-level behaviours in agent-based simulations. These can differ in both scope and resolution. The identification of complex event types gives us a computational method for testing hypotheses about such behaviours, making simulations less 'opaque'. We have demonstrated this by showing correlation relationships between global system behaviours and the interactive mechanisms at lower levels. By showing that correlation relationships can differ amongst different simulations, we have also shown that the same global system behaviour can have different underlying mechanisms, even with the same agent-based model. These multi-level interactive mechanisms are well worth investigating if we are to achieve an understanding of the system beyond simple individual-rule to global behaviour mapping.

Given that complex event types are composed of simple event types (which can be related directly to the agent rules), we have a means of determining which agent rules play a significant role in generating a particular higher level behaviour. Although this has not been discussed in detail in this paper, it is well worth pursuing, particularly if we wish to understand how interventions in a system can affect behaviour. Another promising avenue for further investigation is in the use of more sophisticated statistical methods such as causal state splitting [15] to determine the mechanisms that are critical for a particular higher level behaviour. The complex event formalism would allow us to apply such techniques to behaviours at any level.

Bibliography

[1] BOSCHETTI, F., and R. GRAY, "Emergence and computability", *Emergence: Complexity and Organisation* (2007), 120–130.

[2] BULLEN, T. F., S. FORREST, F. CAMPBELL, A. R. DODSON, M. J. HERSHMAN, D. M. PRITCHARD, J. R. TURNER, M. H. MONTROSE, and A. J. M. WATSON, "Characterization of epithelial cell shedding from human small intestine", *Laboratory Investigation* **86** (2006), 1052–1063.

[3] CHEN, C-C., S. B. NAGL, and C. D. CLACK, "A calculus for multi-level emergent behaviours in component-based systems and simulations", *Proceedings of the satellite conference on Emergent Properties in Artificial and Natural Systems (EPNACS)* (M. A. AZIZ-ALAOUI, C. BERTELLE, M. COSAFTIS, AND G. H. DUCHAMP eds.), (October 2007).

[4] CHEN, C-C., S. B. NAGL, and C. D. CLACK, "Specifying, detecting and analysing emergent behaviours in multi-level agent-based simulations", *Proceedings of the Summer Simulation Conference, Agent-directed simulation,* SCS (2007).

[5] CRUTCHFIELD, J. P., "The calculi of emergence: Computation, dynamics, and induction", *Physica D* **75** (1994), 11–54.

[6] DEGUET, J., Y. DEMAZEAU, and L. MAGNIN, "Elements about the emergence issue: A survey of emergence definitions", *ComPlexUs* **3** (August 2006), 24–31.

[7] DHAR, P. K., H. ZHU, and S. K. MISHRA, "Computational approach to systems biology: From fraction to integration and beyond", *IEEE Transactions on NanoBioscience* **3**, 3 (2004).

[8] KITANO, H., "Computational systems biology", *Nature* **420**, 6912 (November 2002), 206–210.

[9] LAWVERE, F. W., and S. H. SCHAMIEL, *Conceptual mathematics: a first introduction to categories,* Cambridge University Press (1997).

[10] MENG, T. C., S. SOMANI, and P. DHAR, "Modelling and simulation of biological systems with stochasticity", *In silico Biology* **4**, 0024 (2004), 137–158.

[11] MERLO, L. M. F., J. W. PEPPER, B. J. REID, and C. C. MALEY, "Cancer as an evolutionary and ecological process", *Nature Reviews: Cancer* **6** (December 2006), 924–935.

[12] PAOLO, E. A. Di, J. NOBLE, and S. BULLOCK, "Simulation models as opaque thought experiments", *Artificial Life VII: The Seventh International Conference on the Simulation and Synthesis of Living Systems* (Reed College, Portland, Oregon, USA,), (August 2000).

[13] PATEL, M., and S. NAGL, *Cancer Bioinformatics: From Therapy Design to Treatment*, Wiley, (January 2006), ch. Mathematical models of Cancer.

[14] POTTEN, C. S., "Stem cells in gastronintestinal epithelium: numbers, characteristics and death", *Philos. Trans. R. Soc. Lond. B* **353** (1998), 821–830.

[15] SHALIZI, C., *Causal Architecture, Complexity and Self-Organization in Time Series and Cellular Automata*, PhD thesis University of Michigan (2001).

[16] SHALIZI, C. R., *Methods and Techniques of Complex Systems Science: An Overview*, Springer, New York (2006), ch. Methods and Techniques of Complex Systems Science: An Overview, pp. 33–114.

[17] WALKER, D. C., J. SOUTHGATE, G. HILL, M. HALCOMBE, D. R. HOSE, S. M. WOOD, Mac NEIL, and R. H. SMALLWOOD, "The epitheliome: Agent-based modelling of the social behaviour of cells", *BioSystems* **76** (2004), 89–100.

[18] WRIGHT, N., and M. ALISON, *The Biology of Epithelial Cell Populations*, Clarendon, Oxford (1984).

Chapter 3

Developing a complex approach to health phenomena (step 1)

Myriam Patricia Cifuentes
M.D. PhD. Public Health
Universidad Nacional de Colombia
mpcifuentesg@unal.edu.co

Health is a complex object for science and operative levels, partly because there are many approaches defining it but not scientifically sufficient or operatively accepted. This is relevant for health understanding but also for decision making on health related problems. "Determinants of Health" as a widely accepted theoretical proposal, identifies as problematic the reductionist view of health as the disease opposite, attempting to develop it positively according to WHO's definition, proposing a set of factors determining health outcomes. Though this allows a larger comprehension of health causes and effects, still has insufficiently defined theoretical statements and unproved assumptions which difficult understanding and effective actions orientation. Complexity deductive modeling since the insufficiently formalized frameworks, implies incorporating unmanageable object assumptions or reducing health broadness. Taking profit of Bogotá government adherence to DH proposal leading a health information system development, was possible inductive modeling since a systemic massive database (690.000 registries). In this way, DH theoretical statements about health components connectedness were explored by classic statistic approach, and by learning Bayesian networks from data (data mining). First approach showed understanding difficulties. Second was advantageous in approximating within and between determinants relationship structure. However, though DH introduces a systemic approach in considering diverse interacting elements is not empirically satisfactory to exhibit all the meaning of health complexity, because just matches analytic fashioned constructs depending on data expression. A strong networked model developing health complexity, needs the orientation by theoretical constructs as human agency and organization, to explore and understand emergent patterns of health.

1 Introduction

Health is a strongly naturalized and traditional issue in human and societal development, becoming a structural but paradoxically, irresolute and almost unconscious matter, needless to be understood in a deeper way because everybody's roughly knowledge about it seems to be enough for social functioning. According to this, the most popular and widely spread is the easiest "negative notion" of health as disease opposite [1]. Unfortunately this simple notion does not work well while confronting reality manifestations, full of much more dimensions having winding courses and out of control outcomes at many situations, overflowing the disease referent. While explaining infirmity at individual small scale, technical expert diagnosis based upon disease and health exclusive states is blurred when at every day people's life both could coexist in asymptomatic states or adaptive performance. Then additional terms to explain ill health, as sickness or illness must be considered, but also at positive pole other expressions are developed to describe whole life health, as well-being and quality of life. Meanwhile at the large scale, health also shows intricacies, related to same good-bad polarity diagnosis, but evident by unsuccessful interventions with "out of course" results. For example, in developed countries, most expensive technology in health services does not warrant a better people's health, and could be seen that improvement and development in reality coexist with severe poverty related diseases reemergence. For both the "invisible hand of market" rule, does not extrapolate richness to individual nor general wellbeing.

This challenging landscape leads to establish health as scientific object, beyond the biological knowledge, predominant at medicine, towards comprehending its hypothetical complexity [2, 3] contained into reality confusion. In this way, several scientific and academic approaches have been developed, as a basis to transform problematic field with knowledge based decision making, but predominantly focused in just fractions and/or according to biased and analytic points of view.

Theory and prior knowledge is important for complex systems science approach because usually is the point of departure for exploring complex structure and dynamic behavior by means of the predominant methodology of agent based computational simulations. However, health related low and high level human performance is not the focus of current theories statements, because health is not understood as predicative but a quality substantive. Every way, existing theorization must be explored in order to identify its accuracy for the development of a complex systems approach of health phenomenon. Then, developing a scientific complex approach to health is a huge task that involves diverse interconnected steps for integrating at least three broad challenges in a coherent way [4]: Intrinsic or individual character of ill defined health, but also extrinsic astonishingly connected social field with related epistemic diversity. At step 0 was possible to explore several theoretical rational and experience based approaches to health, trough an epistemic diversity grid. Notwithstanding, at theoretical richness discovery that seems to be not conciliating, there was possible to find some conceptual patterns. Heath nature give the idea of being a human made object, flexible enough to fit diverse epistemic needs highly related to human life with the specific attempt to intervene it for modifying unsatisfactory issues. In this way one of those issues is disease, maybe one of strongest and unavoidable "attractor acting" issue, also at theoretical proposals attempting to keep away of it. However, also there is a broad amount of other obligatory human circumstances visible in health, not just located at an adjacent place. In this sense, Determinants of Health proposal [1, 5] is an important present acknowledged but with ancient roots theoretical approach developed since almost five decades. At this paper is presented the step one on developing a complex approach to the health process, concerning complexity exploration of DH theoretical established proposals by means of searching for a complex like structure, looking at the evidence for identifying empirical patterns of co-occurrence [6].

2 Determinants of Health

Determinants of Health proposal represents an important advancement in enlarging the health scope framework, that constitute some kind of hypothetic deductive system evidence supported, which main axe is the statement that health is related to a set of essential factors by a determination link [5, 7]. In this way, there are three recognizable parts in the proposal. First one is health, a complex phenomenon, narrow or social understood [3], but still, ill defined [8, 9], because although considered as a positive matter or special good constitutive of person well being, enabling people to function as agents [3], sometimes is defined in terms of its determinants (as individual capacity [2]) or documented trough known "negative" indicators of ill-health (mortality, morbidity, disability).

Second element is the set of determinants. Four elements were included originally: Environment, lifestyle, human biology and health care systems. The DH general structure success produced an expansive evolution of the idea, and then many approaches of a diverse kind of determinants arrangement were developed, according to multiple points of view involved. Some examples are "Population health" proposal [2, 8, 10] maintaining the same main axe with twelve factors considered, but criticized for being incomplete in important issues [11] as politics; World Health Organization, WHO [12], adheres to "social determinants" emphasis proposal [13, 14] which sub classifies determinants on context, structural and immediate ones, under a determinants unequal relevance assumption.

Third element is about relationships between health and determinants, and within the last. General statement locates factors determination over health, but also existence of complex interactions between determinants. However, it is not a clear sense on an interaction connotation, as an additive term into a model with "independent variables", or if determinants are acting capable over the others. Nevertheless, two way direction interconnections among determinants have been explored and accumulated since a lot of partial evidence, including health indicators, which play a twofold role as determined, but also as determinants. It is not clear if this is a circularity problem on definitions, or if it corresponds to a feedback additional assumption. Although mentioned, social relations role is marginally developed.

2.1 Determinants of Health structural complexity exploration

Recognizing worldwide relevance of DH proposal but also as complexity direct proclaiming one [2, 15], present work step 1 is dedicated to explore the DH complexity coherence character, under the assumption of health as a phenomenon compound by highly interconnected factors. The work is not focused on thematic specificity of particular determinants models [11] whose high variability is context dependent, based upon time-historical, spatio-geopolitical, and epistemic (scientific or not) specificities.

Without predicative agent oriented theoretical statements, deductive approach could not been developed (ie MAS) at this step. By fortune, was possible to have an alternative inductive approach, thanks to data availability collected according to DH proposal, which oriented a special government public health program "Salud a su Hogar", of Bogota District Secretariat of Health. A subpopulation of 100.050 registries was exploited, defined by geopolitical limits and socioeconomic classification of vulnerability, part of a whole data base of 690.000 rows with 139 variables. By cause of whole access to subpopulation data, but also in order to preserve social networked character [6], probabilistic sampling was not done.

1. Classical data approach: The subpopulation database was explored by non parametric first order relation coefficient phi r_o [16] for all variables. Then 9591 different values from a dissimilarity matrix were obtained and explored excluding diagonal ones. Found range of phi values were between 0 and 1, meaning no relation to a total one, but with centrality measures (mean 0.74, median 0.73 and mode 0.98), and dispersion (st.deviation 0.17) showing a right biased distribution, accounting for the strength of relationships established. Highly significance related Xi^2 based values were found, though present in a wide range ($1.39e^{-103}$ to 1), evidenced by centrality and dispersion measures showing a left biased distribution (mean 0.002, median $1.63e^{-18}$ and mode 0.02, st. deviation 0.02) (graphic 1). Then more than 90% phi values were above 0.5, with 90% of significance values below 0.005 cut point, providing evidence for strong general relationships existence, at least of first order.

However, particular cases and graphical examination (graphic 2) let see that extreme upper phi values have less significance values, and also that graphical pattern formation by descendent ordering seems to have sensibility to highest frequency values. This finding is confirmed by Kappa [16] test between ordered correlations and frequencies, significative for 90% of variables. Graphical exploration (graphic 2) also reveals poor organized correlations between and within determinants sets, if compared with an expected pattern according to DH social theory subclassification. Further exploration of general and specific hypothesis or conclusions of fragmented evidence already collected, is increasingly difficult by massive simultaneous outputs of higher order possible relations.

2. Data mining approach: Bayesian Networks (BN) are broadly developed knowledge discovering tools coupling graph and probability theories, to deal with uncertainty and complexity problems [17]. BN are compound by a Directed Acyclic Graph (DAG) and an associated joint probability distribution expressed by means of conditional probability tables (CPT). DAG is constituted by nodes usually corresponding to variables, and by arcs defining conditional or causal dependencies between the nodes. Probability defines quantitatively local relations and general structure by mean of conditional independence simplifying rule.

BN arrange, allows introducing qualitative and quantitative statements of previous cumulated knowledge in a natural human comprehensive way of interconnected expressions, but also provides an interface to external data introduction, as posterior reality observations [17]. Then, prior assumptions and posterior features of a model could be developed. At BN field, there are diverse developed algorithms with different learning capabilities including parameter fitting and structure discovery at same time or separately.

Inductive connectivity exploration of DH proposal by BN learning structure methods, were applied for the same subpopulation data base over 134 variables (excluding those carrying redundant information), without structure assumptions other than those of variable arrange done at the subpopulation survey, by means of software *GeNIe* modeling environment developed by the Decision Systems Laboratory of the University of Pittsburgh (http://dsl.sis.pitt.edu). Reached structure is shown at graphic 3. In general, a reduced 0.75% structure of 497 arcs (Avg indegree: 3.709, Max indegree: 7 –fixed-, Avg outcomes: 3.44, Max outcomes: 21) was obtained instead of a 17889 relationships full connected previous described one. Distances (error test) between real and model generated data percent frequencies was 0,25% on average (median 0.08%, range $5,6-4,9e^{-8}$%, right biased), with F test values account for non significant differences, though over fitting is avoided by intrinsic Occam razor [18].

At general landscape a modest clustered pattern between similar DH color-coded variable become visible. At finer local level of individual variables, a better connectivity fit to intuitive and preceding evidence could be observed. However, neither hierarchical nor thematic organization assumptions seem to consistently emerge. In order to diminish the still high connectivity confusing an easy view, middle level of 13 subsystems, according to general DH thematic, were organized. This arrange produces an easier view, but hierarchy still does not emerge, because many of subsystems as also single variables, act as similar weighted hubs (high connectivity nodes) that could indicate the need of intermediate ones that could be hidden [17]. One of those subsystems is "heath indicators" one, as receiver and also source, with a 76% connectivity.

3. Discussion

Bayesian networks seem to be important tools for modeling or constructing reality based complex systems, widely developed at field of artificial intelligence data mining techniques. Although BN (and other now available data mining tools) are able

31

of fitting massive data hidden structure, assumption about empirical neutrality and ability of lower level granularity (data) as paramount way to know the world, becomes insufficient, even though the broad access allowed at actual information, communication and knowledge shaped societies. At the other side rational deductive development from theories to models tested against reality does not take full profit of evidence potential. Every way we cannot forget that data have roots on cognitive human points of view about reality, as also have human theorization mechanisms. In this way, neither increasing ability to record every human and natural phenomenon or better tools for deductive reason, are isolated warranties of better reality comprehension.

At this paper an already but ill fashioned wide accepted theory for health, was tested by unsupervised BN approach taking profit of a big amount of data representing a governmental interpretation for measuring all the fragments on DH. At a different granularity levels some degree of emergent organization was observed, but not consistently enough with intuitive reasoning and previous partial evidence results. This could be produced but insufficiently theoretic definition or still insufficient data, or both. Data insufficiency was an old problem and perhaps ever will be, though increasing sources availability, by cause of the always lack of fitness with changing needs. But theory accounts for universals, that could be changed or actualized, lasting a more long time period, robust to contingencies. In this way DH statements about multiple interconnected elements did not perform adequately for a complex system perspective. Then better theoretical approaches are needed, including rational but also evidence based for its development.

In order to it, subsequently steps must be accomplished. Former one is about developing health theoretical statements (blind point) according to complex systems science multiagent framework, as way to recover the human and predicative character of health main axe, instead of occasion, unstable, highly artificial and interest oriented topics as DH develops. This approach could have the advantage of being the seed for taking profit of the natural multileveled organized human structures, which let to solve individual-social dichotomies. This step must go along with evidence incorporation. Latter step based upon multiagent networked health structure, is about modeling its dynamic behavior, according to human lifecycle dynamics. Additional steps beyond, since general structural and dynamic rules identified, are on simulation environments development which allows universalities improvement but also robust prediction tools creation utilizable at research, information systems improvement, policy and decision making and education.

Bibliography

[1] Davidhizar, R., 1983, Critique of the health-belief model. Journal of Advanced Nursing, 8:467-472.

[2] Public Health Agency of Canada. 2007, Towards a Common Understanding: Clarifying the Core Concepts of Population Health. http://www.phac-aspc.gc.ca/ph-sp/phdd/docs/common/index.html

[3] Solar, O., Irwin, A., 2007A conceptual framework for action on the social determinants of health. Commission on Social Determinants of Health, WHO.

[4] National Cancer Institute, 2006, Greater Than the Sum: Systems Thinking in Tobacco Control. NIH Pub. 06-6058.

[5] Glouberman, S., Millar, J., 2003, Evolution of the determinants of Health, Health Policy, and Health Information Systems in Canada. American Journal of Public Health, 93:388-392.

[6] Hanneman, R., 1988, Computer assisted theory building. Modeling dynamic social systems: SAGE Publications.

[7] Bunge, M., 1997 (1959), La causalidad. El Principio de Causalidad en la Ciencia Moderna. Buenos Aires: Editorial Sudamericana.

[8] Evans, RG, Stoddart, G., 2003, Consuming Research, Producing Policy? American Journal of Public Health, 93, 3.

[9] Almeida Filho, N., 2000, O conceito de saude: ponto-cego da epidemiologia? Revista Bras. Epid, 3, 1.

[10] Public Health Agency of Canada., 2007, Population Health. What determines Health? from "Toward a Healthy future". http://www.phac-aspc.gc.ca/ph-sp/phdd/determinants/determinants.html#unhealthy

[11] Evans, R., Barer, M., Mermor, T., 1996, ¿Por qué alguna gente está sana y otra no? Los determinantes de la salud en las poblaciones. Madrid, Ediciones Diaz de Santos, S.A.

[12] Kindig, D., Stoddart, G., 2003, What is Population Health? American Journal of Public Health, 93. 3.

[13] Coburn, D., Denny, K., Mykhalovskiy, E., McDonough, P., Robertson, A., Lowe, R., 2003, Population Health in Canada: A brief Critique. American Journal of Public Health, 93, 3.

[14] Mathers, C., 2004, Towards a conceptualization of health. In: UNECE/WHO/Eurostat Meeting on the Measurement of Health Status, Geneva.

[15] Marmot, M., Wilkinson, R. (Eds), 2000, Social Determinants of Health, Reprint edn. Great Britain: Oxford University Press.

[16] Wilkinson, R., Marmot, M., 2003, The Social Determinants of Health. The Solid Facts. Second Edition, World Health Organization, Europe.

[17] Public Health Agency of Canada, 2007, Towards a Common Understanding: Clarifying the Core Concepts of Population Health. Appendix A. http://www.phac-aspc.gc.ca/ph-sp/phdd/docs/common/index.html

[18] Siegel, S., Castellan, J., 1985, Estadística no paramétrica. Aplicada a las ciencias de la conducta., Editorial Trillas, Mexico D.F.

[19] Murphy, K., 1998, A brief Introduction to Graphical Models and Bayesian Networks.

[20] Myllymäki, P., 2002, On Probabilistic modeling and Bayesian Networks Complex Systems Computation Group. Helsinki Institute for Information Technology, Finland.

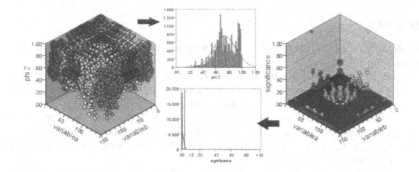

Graphic 1: Phi values 3d plot and histogram at left side. At right side are significance values 3D plot and histogram, with power 0.5 scale transformation to improve visualization, and 0.05 and 0.005 cut points. 3D plot color code automatically assigned is kept to allow visualization.

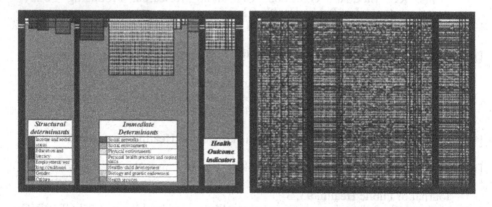

Graphic 2: Determinants class internal expected coherence (left) and observed (right) graphic patterns by descendent ordering of phi values by groups of determinants color defined at first line in both, with color codes at first graphic bottom.

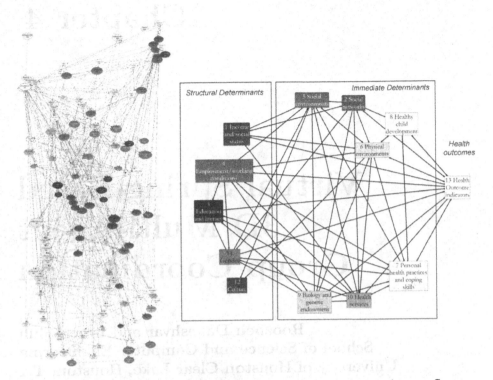

Graphic 3: General BN structure (left), where nodes are variables and arcs reflect relationships between them. Same color codes are maintained. At right side middle level BN structure arranged by subsystems and them 3 hierarchies. GENIE Software lets exploring variables inside subsystems.

Chapter 4

Virtual Spring-Based 3D Multi-Agent Group Coordination

Roozbeh Daneshvar and Liwen Shih
School of Science and Computer Engineering
University of Houston Clear Lake, Houston, TX
roozbeh@tamu.edu, shih@uhcl.edu

As future personal vehicles start enjoying the ability to fly, tackling safe transportation coordination can be a tremendous task, far beyond the current challenge on radar screen monitoring of the already saturated air traffic control. Our focus is on the distributed safe-distance coordination among a group of autonomous flying vehicle agents, where each follows its own current straight-line direction in a 3D space with variable speeds. A virtual spring-based model is proposed for the group coordination. Within a specified neighborhood radius, each vehicle forms a virtual connection with each neighbor vehicle by a virtual spring. As the vehicle changes its position, speed and altitude, the total resultant forces on each virtual spring try to maintain zero by moving to the mechanical equilibrium point. The agents then add the simple total virtual spring constraints to their movements to determine their next positions individually. Together, the multi-agent vehicles reach a group behavior, where each of them keeps a minimal safe-distance with others. A new safe behavior thus arises in the group level. With the proposed virtual spring coordination model, the vehicles need no direct communication with each other, require only minimum local processing resources, and the control is completely distributed. New behaviors can now be formulated and studied based on the proposed model, e.g., how a fast driving vehicle can find its way though the crowd by avoiding the other vehicles effortlessly[1].

1 Introduction

Multi agent systems offer many potential advantages with respect to single-agent systems such as speedup in task execution, robustness with respect to failure of one or more agents, and scalability [17]. The role of larger multi-agent systems has become more significant in recent years, due to lower cost for simpler agents and increased potential group capabilities in robustness and flexibility. These newly evolved, highly complex large-scaled multi-agent systems demand improved interaction study and innovated group coordination approaches. Despite these initial efforts, further investigations are desperately needed in this new group emergence paradigm. In the following, we will first review the current progress in multi-agent coordination field, and then propose a Virtual-Spring based group coordination to cope with the increased complexity, as the problem scaled to many more agents in 3D.

1.1 Flying Vehicles

A multi-agent System for formation flying missions is proposed in [19] and for collaborative sensing, multiple Unmanned Aerial Vehicles (UAVs) are considered in [18].Methods for optimizing the task allocation problem for a fleet of UAVs with tightly coupled tasks and rigid relative timing constraints are described in [2]. Minimization of the mission completion time for the fleet is the overall objective in this work that uses timing constraints and loitering. The problem of decentralized task assignment for a fleet of cooperative UAVs is considered in [1] which extends the analysis of the algorithm of previous work to consider the performance with different communication network topologies. In [9], [7] and [10] cooperative UAV routing with limited sensor range is considered (the problem for one UAV is investigated in [8]).

1.2 Spatial Multi Agent Systems and their Coordination

Multi agent coordination techniques are used in various tasks like Air Traffic Management [16] while there are several decentralized algorithms like [13] for aircraft-like vehicles. Air Traffic Management of the future allows for the possibility of free flight, in which aircrafts choose their own optimal routes, altitudes, and velocities. The safe resolution of trajectory conflicts between aircraft is necessary to the success of such a distributed control system [21]. In [14] it is tried to capture the idea that the less coordination a multi-robot system requires, the better it should scale to large numbers of robots. In [6] the real-time multi-agent coordination and control requirements of automobile and submarine systems are discussed. The use of hybrid systems techniques for analyzing and synthesizing

---

[1] Any kind of military uses from the content and approaches of this article is against the intent of the authors.

the control architectures have been under investigation. Also in [12] a method for cooperative control in a distributed autonomous robotic system is proposed. A reactive navigation strategy is used for controllers of robots by combining repulsion from obstacles with attraction to a goal. A class of dynamic vehicle routing problems (in which a number of mobile agents in the plane must visit target points generated over time by a stochastic process) are considered in [4]. The aim has been to minimize the expected time between the appearance of a target point and the time it is visited by one of the agents by making minimal or no assumptions on communications between agents. It is shown that inter-agent communication does not improve the efficiency of such systems, but merely affects the rate of convergence to the steady state. In [17] a policy for steering multiple vehicles between assigned independent start and goal configurations is proposed which ensures collision avoidance. The decentralized policy rests on the assumption that agents are all cooperating by implementing the same traffic rules (each agent decides its own motion by applying those rules only on locally available information). In [5], formations of robots are considered. In this work a motion plan for the overall formation is used to control a single leader and the followers use local control laws.

A simulation environment for massive systems is proposed in [3] which is capable of coping with 3D environments. In [15] an approach to qualitative spatial orientation reasoning in 3-dimensional spatial environment is proposed. The problem of positioning a group of autonomous but coordinating mobile robots into a specified spatial configuration is considered in [11]. In this work there is no central controller or inter-agent communication. The robots move into position without collision or unnecessary delay. A hierarchical controller with three levels (Execution, Coordination and Organization) has been adopted in their approach.

2 Spatial Coordination using Spring Forces

A virtual street is built for flying vehicles as shown in Fig.1. This environment is made with NetLogo program [22]. In this program a setup code specifies the initial positions and orientations. The group of agents moves in a 3D space and the behavior of each agent for every time step is defined by a piece of code.

One of the approaches is to use virtual springs for coordination of agents. This idea has been used for coordination of soccer simulation footballers in [20] in which each of the players was constrained with a group of springs with other teammates. As a result, the team demonstrated group behaviors that met the desired criteria.

An agent is constrained by a group of springs that specify the later position of the agent in the space (This applies similarly to both 2D and 3D environments). The length, stiffness and the total number of the springs can vary from agent to agent depending on the environments, systems, and designers. Under a group of forces, an agent moves until the total resultant forces (vectors) becomes zero on the agent. When the spring forces are applied to an agent, the total applied

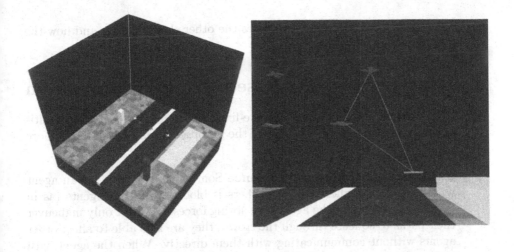

Figure 1: The virtual street with some flying vehicles in the two sides (left) and spatial coordination of vehicles (right)

force is calculated with Eq.1.

$$F_x = \sum_{i=1}^{n} \frac{\Delta L_i \times K_i \times (X_i - X_A)}{D_i}$$

$$F_y = \sum_{i=1}^{n} \frac{\Delta L_i \times K_i \times (Y_i - Y_A)}{D_i}$$

$$F_z = \sum_{i=1}^{n} \frac{\Delta L_i \times K_i \times (Z_i - Z_A)}{D_i} \tag{1}$$

in which X_A, Y_A and Z_A specify present position of agent, n is the number of factors the agent has spring connections with (for this case it is equal to the number of neighbors), L_i is the length of the spring with i^{th} factor, K_i is the constant of that spring, D_i is the distance to the ith factor and (X_i, Y_i) specify the position of the ith point.

Hence the agent moves towards the direction of the applied force to get closer to the mechanical equilibrium point. This changes the position of the agent and is able to play a role in the coordination of the group (the agent has only observed the positions of the neighbors). The proposed spring-based model is dimension scalable, where no extra effort is needed for agents to make movement decisions extending from 2D to 3 D space (see Fig.1).

Agents with different velocities show different behaviors for finding their paths (according to their relative velocity to other members of the group). As an example, Fig.2 illustrates how the fastest agent overtakes the other slow agents and how the inter-agent distances change in time. It shows that how the agent driving faster than others, overtakes the other slower agents and how the distances change during time.

3   Benefits of Spring-Based Group Coordination

Compared with the current, generally pre-fixed and centrally controlled multi-agents group coordination, the benefits of the proposed spring-based method are detailed as follows:

- **Swift Mutual Collision Avoidance** Sometimes the path for an agent with a velocity higher than the others is blocked by front agents (as in Fig.3). In this case if the agents use spring forces, they not only maneuver their paths to squeeze through the crowd, they are also able to affect other agents without communicating with them directly. When the agent with higher velocity gets too close to other neighbors, they change to yield their positions respectively (to move to the new mechanical equilibrium point) and hence new space for the agent is made so that it can overtake.

- **Simple New Agents Inclusion to the Group** When new vehicles want to join the group of agents, by the spring method they can enter the group just by setting new positions in the space (probably in the group) as their set points. If their positions were fixed in the group, a new vehicle should have waited for an available empty space to join the 3D street.

- **Easy 3D Maneuver** When a vehicle wants to move horizontally or vertically (X or Z directions, if considering the direction of street as Y) it only needs to move to that direction slowly while obeying the same coordination process. This forces other neighbors to leave more room for the vehicle to change its position. A sample is shown in Fig.2 in which the agent is moving to the right constantly and for some periods it has to find its path (for demonstration purposes, the agent is moved to the left most position when it reaches the right most limit).

4   Conclusion

In this article, we investigated a Virtual-Spring based group coordination model in 3D environments. The main aspect considered in this model was to demonstrate how collision-avoidance group level behaviors emerge from the interactions of agents with their individual safe-distance minded behaviors.

Bibliography

[1] ALIGHANBARI, Mehdi, and Jonathan P. HOW, "Robust decentralized task assignment for cooperative uavs", *AIAA Guidance, Navigation and Control Conference*, (2006).

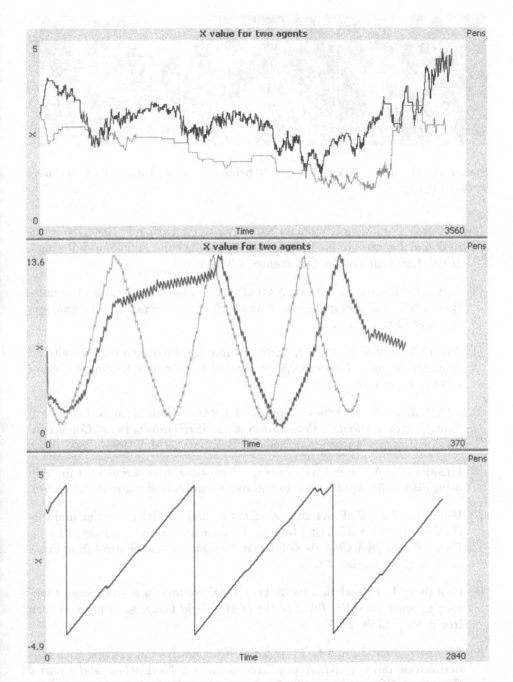

Figure 2: Distances of two normal agents from the agent which has the most velocity in the group (top), Distances of two normal agents from the agent which has the most velocity in the group (middle) and the X value for an agent which drives to the right at each step beside doing coordination (bottom)

41

Figure 3: The agent with white color is behind the other three agents who have blocked the way

[2] ALIGHANBARI, Mehdi, Yoshiaki KUWATA, and Jonathan P. HOW, "Coordination and control of multiple uavs with timing constraints and loitering", *IEEE American Control Conference*, (June 2003).

[3] AOYAGI, Masaru, and Akira NAMATAME, "Massive multi-agent simulation in 3d", *Soft Computing as Transdisciplinary Science and Technology*, Springer (2005), 260–272.

[4] ARSIE, A., and E. FRAZZOLI, "Efficient routing of multiple vehicles with no communications", *International Journal of Robust and Nonlinear Control* (2006), To appear.

[5] DESAI, J. P., J. OSTROWSKI, and V. KUMAR, "Controlling formations of multiple mobile robots", *Proceedings of the IEEE International Conference on Robotics and Automation* (Leuven, Belgium,), (May 1998), 2864–2869.

[6] DESHPANDE, A., and J. de SOUSA, "Real-time multi-agent coordination using diadem 5 : Applications to automobile and submarine control" (1997).

[7] ENRIGHT, J.J., E. FRAZZOLI, K. SAVLA, and F. BULLO, "On multiple UAV routing with stochastic targets: Performance bounds and algorithms", *Proc. of the AIAA Conf. on Guidance, Navigation, and Control* (San Francisco, CA,), (August 2005).

[8] ENRIGHT, J. J., and E. FRAZZOLI, "UAV routing in a stochastic, time-varying environment", *Proc. of the IFAC World Congress* (Prague, Czech Republic,), (July 2005).

[9] ENRIGHT, John J., and E. FRAZZOLI, "Cooperative uav routing with limited sensor range", *AIAA Conf. on Guidance, Navigation, and Control* (Keystone, CO,), (August 2006), Paper AIAA-2006-6208.

[10] FRAZZOLI, E., "Maneuver-based motion planning and coordination for multiple UAVs", *Proc. of the AIAA/IEEE Digital Avionics Systems Conference* (Irvine, CA,), (2002).

[11] GOLD, Timothy B., James K. ARCHIBALD, and Richard L. FROST, "A utility approach to multi-agent coordination", *IEEE International Conference on Robotics and Automation* (San Francisco, CA,), (April 2000).

[12] HANG, Seong-Woo, Shang-Woon SHIN, and Doo-Sung AHN, "Formation control based on artificial intelligence for multi-agent coordination", *Proceedings of the IEEE ISIE 2001* (Pusan, Korea,), (2001).

[13] INALHAN, Gokhan, Dusan STIPANOVIC, and Claire J. TOMLIN, "Decentralized optimization, with application to multiple aircraft coordination", *41st IEEE Conference on Decision and Control* (Las Vegas, NV,), IEEE, (December 2002).

[14] KLAVINS, Eric, "Communication complexity of multi-robot systems", *Algorithmic Foundations of Robotics V*, (J.-D. BOISSONNAT, J. BURDICK, K. GOLDBERG, AND S. HUTCHINSON eds.) vol. 7 of *Springer Tracts in Advanced Robotics*. Springer (2003), pp. 275–292.

[15] LIU, Jianhui, "Qualitative orientation reasoning in spatial multi-agent environment.", *IAT*, (2005), 266–272.

[16] NGUYEN-DUC, Minh, Jean-Pierre BRIOT, and Alexis DROGOUL, "An application of multi-agent coordination techniques in air traffic management", *Proceedings of the IEEE/WIC International Conference on Intelligent Agent Technology (IAT03)*, (2003).

[17] PALLOTTINO, L., V.G. SCORDIO, E. FRAZZOLI, and A. BICCHI, "Decentralized cooperative policy for conflict resolution in multi-vehicle systems", *IEEE Trans. on Robotics* (2007), To appear.

[18] PARUNAK, H. Van Dyke, Sven A. BRUECKNER, and James J. ODELL, "Swarming coordination of multiple uavs for collaborative sensing", *Second 2ND AIAA Unmanned Unlimited Systems Technologies and Operations Aerospace Land and Sea Conference and Workshop and Exhibit* (San Diego, CA,), (September 2003).

[19] SANDA MANDUTIANU, Fred Hadaegh, and Paul ELLIOT, "Multi-agent system for formation flying missions".

[20] SHARBAFI, Maziar, Roozbeh DANESHVAR, and Caro LUCAS, "New implicit approach in coordination between soccer simulation footballers", *Fifth Conference on Computer Science and Information Technologies (CSIT 2005)* (Yerevan, Armenia,), (September 2005).

[21] TOMLIN, Claire J., George J. PAPPAS, and Shankar SASTRY, "Conflict resolution for air traffic management: A study in multi-agent hybrid systems", *IEEE Transactions on Automatic Control* **43**, 4 (April 1998).

[22] WILENSKY, Uri, "NetLogo: Center for connected learning and computer-based modeling, Northwestern University. Evanston, IL" (1999).

Chapter 5

Exploration for Agents with Different Personalities in Unknown Environments

Sarjoun Doumit and Ali Minai

Complex Adaptive Systems Laboratory (**casl**)
University of Cincinnati, Ohio, U.S.A.
doumitss@email.uc.edu, Ali.Minai@uc.edu

We present in this paper a personality-based architecture (PA) that combines elements from the subsumption architecture and reinforcement learning to find alternate solutions for problems facing artificial agents exploring unknown environments. The underlying PA algorithm is decomposed into layers according to the different (non-contiguous) stages that our agent passes in, which in turn are influenced by the sources of rewards present in the environment. The cumulative rewards collected by an agent, in addition to its internal composition serve as factors in shaping its personality. In missions where multiple agents are deployed, our solution-goal is to allow each of the agents develop its own distinct personality in order for the collective to reach a balanced society, which then can accumulate the largest possible amount of rewards for the agent and society as well. The architecture is tested in a simulated matrix world which embodies different types of positive rewards and negative rewards. Varying experiments are performed to compare the performance of our algorithm with other algorithms under the same environment conditions. The use of our architecture accelerates the overall adaptation of the agents to their environment and goals by allowing the emergence of an optimal society of agents with different personalities. We believe that our approach achieves much efficient results when compared to other more restrictive policy designs.

1 Introduction

Rodney Brooks' subsumption architecture(SA)[1] was meant as a reactive system for an exploratory robot that kept no memory or stored information from its environment. When applied in real life, the result was a robot that reacted to its surrounding's input and adjusted its actions accordingly, and while it did survive its environment, it did not store any meaningful abstraction of its experiences. Because the subsumption architecture (SA) decomposes a system into parallel tasks (or behaviors) of increasing layers of competence, (versus the typical functional decomposition), it allows the system to grow incrementally and become resilient to sudden and unexpected changes in the environment. But on the other hand, the SA lacks the flexibility to implement higher-level concepts which allows it to be more useful for a wider scope of applications. For example, application involving software-based agents and robots with hybrid software/hardware controllers domains find it very difficult to apply a pure subsumption architecture to their design [2][3][4]. Our goal is to allow our agents to manipulate at a reflexive level, the knowledge that the environment presents and handle more complex and challenging missions by introducing hybrid learning algorithm (at a higher level. by introducing *personalities*). In our system, we consider the changing or driving forces that impacts the personality of the agent to be the activity of seeking different types of rewards from its environment and how this impacts the agent in choosing its best fitting personality. Using this paradigm, the design of our algorithm challenged us to address a fundamental issue in psychology, which is nature vs. nurture. When designing an agent, how much information should we embed into the agent and how much should we let the environment shape it on its own? We believe we achieved a balanced approach to solving this issue through our PA architecture. In the next section we discuss our model, followed by the experiment environments and the results.

2 The PA model

We present the schematic diagram in figure 1 of our PA architecture by describing our *behavior*'s architecture, and then the personality's model.

2.1 The behavior architecture

We define a behavior to be a collection of *actions*. An *action* represents a procedural heuristic of several small processes that an agent can perform at a reflexive level. In relevance to Brooke's subsumption architecture, where we have different layers of behaviors, a behavior in our architecture can comprise multiple layers of actions. All actions receive input from the environment and forward their outputs to *junctions* via *connections*. A *junction* serves as an evaluator/relay for all its "inputs" and then outputs the resulting evaluation. In a simple behavior, shown in the diagram to the left in figure 1, we see a behavior comprised of 2 actions: Action A and Action B with a single junction

`Junction J`, where the junction's output represents the overall behavior output. In more elaborate behaviors, we have layers of actions, such as the composite behavior to the right in figure 1. In these behaviors, the junctions evaluate the actions across the same level and also, from junctions that forward their outputs to other junctions, i.e. `Junction J` in level 1 subsumes/connects to `Junction J'` in layer 2 in `C-JJ'`. The final junction in the lowest layer handles the final output, or the final output of the behavior.

$$J = \sum_{i=1}^{n} W_i C_{i,J} + \sum_{k=1}^{m} W_k C_{k,J}$$

where J is the junction. i represents the available actions and k the number of other junctions that input into J. W_i is the weight of action i and $C_{i,J}$ is the weight of the connection from action i to junction J. (similarly for the junctions k). Every individual action and junction has an associated weight that deter-

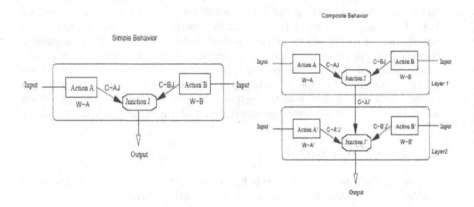

Figure 1: Behavior Architecture.

mines the relevance of its output to the behavior's dynamics, i.e. `Action A` has weight `W-A`. Also, every connection has a weight that works as an attenuator or reinforcer to the output value it represents, i.e. the connection connecting `Action A` to `Junction J` is `C-AJ`. It is worth mentioning also that the layers are not hard boundaries, and that an `action A` in a high level can "subsume" (be connected via a junction to) the output of an `action A'` in a lower level. The behavior's architecture is based on a neural network design with more functionalities added at the neuron level.

2.2 The personality model

In our architecture an agent's *personality*, see figure 2, represents the collective values and weights of a behavior's components (actions, junction, connections). Our personality architecture applies a reinforcement learning architecture where the learning algorithm's objective is to maximize the available rewards present in the environment. This is achieved by predicting what best values for the weights and connections for a behavior to assume that would let the agent collect the largest number of rewards possible. As mentioned earlier, an action's weight determines the action's relevance and impact on the final output of the behavior, similarly connections' weights determine if this connection is valid, strong or non-existent. All these values have the effect of shaping the generic behavior architecture into more specialized behaviors, for example in the `Composite Behavior` to the right in figure 1, if the value of `C-BJ = 0`, then this means that `Action B`, and its contribution, are not part of the architecture. In practice, this change could mean the difference between an agent that has an *aggressive behavior* or a *careful behavior*. The ramifications of having different values for all the involved *parameters* is that we end up having an endless number of possible behaviors. This of course seems to be the natural way of the world, as in relevance to human beings, every person has a distinct personality that allows him/her to behave and respond uniquely to a situation. For simplicity and to facilitate the analysis of the system, we decided to make it a discrete learning problem where we define precisely what constitutes the different kinds of possible behaviors we wish to have. The personality therefore can pick from a *mood* collection (containing the lists of corresponding values and weights), the possible behavior type or *mood* it wishes to be in. The learning algorithm is divided into subtasks to maximize the corresponding reward components. For every type of identifiable reward, we assign a subtask that contributes towards maximizing the cumulative total reward. The PA architecture is especially geared towards multiple reward sources and multiple goal requirements in order to bridge and marry these goals into a uniform behavior that maximizes the overall rewards. In the extreme case where there is one identifiable reward, the agent distinguishes between its *internal state of rewards* and the *external state of rewards*, (which is the internal state of rewards of other agents it comes into contact with), and assigns an *emotional maturity arbitrator* (or M) to each one of them that serves a predictor of the expected rewards. Let S_i, be the current behavior mood of the agent, where i represents its index in an array of available behaviors $[i..n]$ where n is some integer. Let R_i be a reward with index i associating behavior S_i to reward R_i. Also let λ_i be a decreasing factor (constant) for reward value R_i as time progresses and the agent is still in behavior S_i. Then we can write M as the expected sum of potential future rewards for time $t + n$ (where n are discrete time units) as:

$$M_i(S_i) = E(R_{i,t+1} + \lambda_i R_{i,t+2} + \lambda_i R_{i,t+3} + ... \lambda_i R_{i,t+n})$$

$M_i(S_i)$ is the emotional maturity arbitrator value for behavior S_i at time t.

The winning mood (Si) is projected to the template behavior architecture to apply to the corresponding virtual behavior. In a way the template behavior architecture acts as a projector, and every behavior type is a projection or a virtual representation of the template behavior. (To clarify what we meant by stages in the abstract, the moods of the personality are actually the stages). Finally the rewards are abstractions for an accumulated stimulation or incentives. They are accumulated by the agent, and the agent can have either no, partial or full access to the rewards of other agents, depending on the personality of other agents (how much they like to share). The information from other agents allows the agent to assess its performance and stance with respect to the rest.

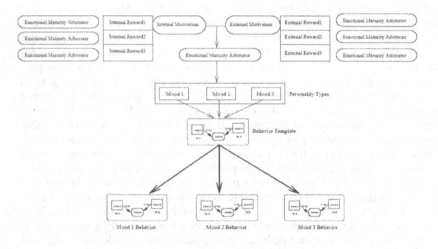

Figure 2: Personality Architecture.

3 Experiments

To test our propositions and validate the utility of our PA, we developed a 3D simulated environment with artificial agents and applied our concepts to the agents in the terrain. We show how by applying the PA to agents with a specific goal allows the emergence of an optimal society of agents with personalities and behavioral dynamics fit to perform that task. For our testing, we chose the simplest and most visited example for this kind of applications, namely exploration by an autonomous agent. We chose this example in particular to i) compare the performance of our architecture and stem against a well understood and studied case, in which the subsumption architecture presents the ultimate solution given its simplicity and specification compatibility, ii) to show how the

PA performs in a situation where there is one main objective (i.e. exploration) which is against the PA's strong points: i.e. joining multiple objectives, iii) limited space to address more elaborate applications such as exploration and exploitation.

3.1 The environment

The environment(terrain) we're describing is a 3 dimensional lattice $(N \times N \times Z)$ where (N) represents the discrete longitude and latitude value and (Z) represents the elevation value at every $(N \times N)$ coordinate, or locations. The value of N ranges discretely from 0 till N. The elevation Z represents an associated cost matrix drawn over the 2 dimensional $(N \times N)$ area (where the cost is a penalty value drawn from the agent's energy and time). The terrain is textured, to present to the agent another type of cost penalty, where every different type of texture represents a different resistance factor μ to the agent's velocity $(\vec{V})$ by $\mu \times \vec{V}$ and energy consumption.

3.2 The agents

Our simulated agents are relatively small robotic vehicles that are capable of limited data processing and communication capabilities. Each agent has a limited non-renewable energy source, and all its functionalities (mobility and communication) draw their energy requirements from it. Every agent is capable of detecting its own remaining energy level and can move in all directions of the compass $(0° \rightarrow 360°)$ at a velocity that is determined by factors such as the terrain's texture and terrain elevation (uphill or downhill).

3.3 The behavior architecture

Our exploration behavior is made up of 5 different actions divided in 3 layers of subsumption dominance, see figure 3. In the top level we have the Avoid Boundaries and Avoid Obstacles actions, in the middle level we have the Wander action and in the low level we have the Sweep and Line actions. As each of their namesakes reveal, every action performs in manners that reflect that sub-behavior. The Avoid Boundaries action allows the agent to steer away areas to avoid crossing into territory that is not part of the mission. Avoid Obstacles on the other hand helps direct an agent around an obstacle (if possible), and allows it to resume afterwards whatever course it was taking, once it cleared the obstacle. Wander is an action for the agent to move randomly in the terrain, Sweep makes the agent systematically explore every location in a specified area, while Line makes the agent explore in a straight line.

We also devised 3 types of moods that affect the behavior architecture's variables shown in figure 3 (i.e. the weights and connections). The moods or personality types are shown in figure 4 and are Blind, Tolerant and Dedicated. In the coming paragraph, the terms high, low, moderate and average represent

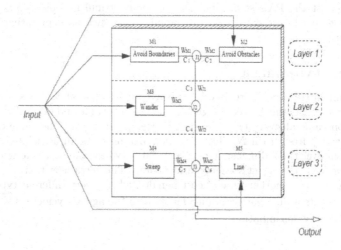

Input

Figure 3: Exploration Behavior Architecture

the values that one might assume for the simulation. They are in reference to each other and do not represent a specific number. In the simulation results we will show the numerical values that we used. The Blind personality is usually characteristic of a fresh agent because it has no rewards of any kind. It does not "worry" much about negative rewards (hard terrain), for its Avoid Obstacles and Avoid Boundaries it has relatively low weights. Its Wander action is extremely random, and its Sweep areas are relatively large and its weights for Sweep are higher than that of the Line. (It appears as if it is in "desperate need" to accumulate rewards no matter what the cost). On the other hand, the Dedicated personality is usually of an agent that has usually accumulated one type of reward more than the other. Its weights for the avoid boundaries and Avoid obstacles is high, while its Wander's direction values are within a short range from each other and its weight is low. Its Sweep's areas are relatively smaller, so its Line's values. The Tolerant personality is usually for an agent

Figure 4: Personality Types (Moods)

that has a balance between its current values for its rewards. It has high weights for its avoid actions, its `Wander`'s direction values range moderately and it has a high weight for it, while it favors the `Line` action over the `Sweep` action for which it has moderate values (area and line). For the purposes of this experiment we made all personalities want to share all the information when nearby agents ask for it. We devised no negative reward for the communication, in order to keep the application simple and focused. Following are the results that we collected.

4 Results

For both our experiments our $N = 100$ and $Z = 0^o, 30^o, 60^o$. Agents move at a constant velocity V of 1 N unit per 1 time unit when at $Z = 0^o$ and texture $\mu = 1$. The agents have a power supply of 10 Joules, and their energy consumption matches that of small robots according to the average from various technical reports. The agents are distributed randomly on the terrain (not for the same experiment) and all start at the same time.

For our first experiment, we ran simulations to compared the average results of 5 agents that employ a pure subsumption architecture vs 5 agents that employ our PA (with its 3 available moods). The agents using PA all started with randomly selected personalities and the simulation ran for x time units. The results are shown in graphs 5 As the results show, the overall performance of the PA is

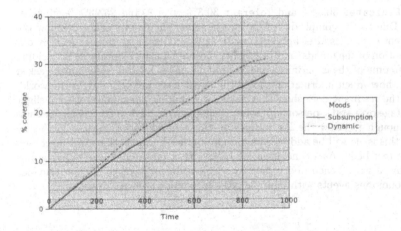

Figure 5: Exploration Coverage Ratio between PA and Subsumption

better than that of a pure subsumption architecture despite the simplicity of the application. Both architectures match up in the first phases of the experiment but as time progresses PA's out performs subsumption.

For our second experiment shown in graph 6, we compared the average results of 5 agents that are always `Blind`, 5 agents that are always `Dedicated`, 5 that are `Tolerant` and 5 that have a dynamics personality. The purpose of this experiment was to compare (on average) the performance of these 4 personalities. The results showed an almost identical match amongst the Dynamic

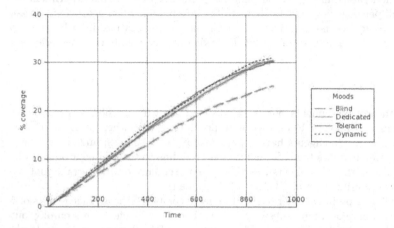

Figure 6: Exploration Coverage Ratio between different static and dynamic personalities

31%, Dedicated 30.54% and Tolerant 30.3% with Blind 25.23% lagging behind. Due to the complexity of the experiments, the results can only show the emergent dynamics after applying our designs and architectures. Obviously our designation of the moods and the weights chosen played a pivotal role in deciding the outcome of the experiment, which brought us back to the question we asked before, how much information should the agent have before going to the world. From the results it does show that having a changing personality offers a slight advantage over static personalities, but then again the task at hand was not complex enough to allow the PA to improve the general performance. For future work, this issue will be addressed and more in depth applications will be chosen for the test beds. As a conclusion, we believe that our architecture offers a first step and a work-bench in the right direction to improve upon the performance of autonomous agents with multiple goals in unknown environments.

Bibliography

[1] BROOKS, Rodney, "Sample technical report", *Tech. Rep. no. 864*, A Robust layered control system for a mobile robot, (Sept. 1985).

[2] FRANKLIN, Stan, and Art GRAESSE, "Is it an agent, or just a program: A taxonomy for autonomous agents", *Tech. Rep. no.*, Institute for Intelligent Systems. University of Memphis, (1996).

[3] ROSENBLATT, J., S. WILLAMS, and H. DURRANT-WHYTE, "Behavior-based control for autonomous underwater exploration", *IEEE International Conference on Robotics and Automation*, (Apr. 2000).

[4] UNIVERSITY, North Carolina State, "Pipe-crawling robots designed to find earthquake, bomb survivors", *Tech. Rep. no.*, North Carolina State University, (Sept. 1999).

Chapter 6
The Self-Made Puzzle:
Integrating Self-Assembly and Pattern Formation Under Non-Random Genetic Regulation

René Doursat
Institut des Systèmes Complexes, CREA
CNRS & Ecole Polytechnique, Paris
http://doursat.free.fr

On the one hand, research in *self-assembling* systems, whether natural or artificial, has traditionally focused on pre-existing components endowed with fixed shapes. Biological development, by contrast, dynamically creates new cells that acquire selective adhesion properties through differentiation induced by their neighborhood. On the other hand, *pattern formation* phenomena are generally construed as orderly states of activity on top of a continuous 2-D or 3-D substrate. Yet, again, the spontaneous patterning of an organism into domains of gene expression arises within a multicellular medium in perpetual expansion and reshaping. Finally, both phenomena are often thought in terms of *stochastic* events, whether mixed components that randomly collide in self-assembly, or spots and stripes that occur unpredictably from instabilities in pattern formation. Here too, these notions need significant revision if they are to be extended and applied to embryogenesis. Cells are not randomly mixed but pre-positioned where cell division occurs. Genetic identity domains are not randomly distributed but highly regulated in number and position. In this work, I present a computational model of *programmable* and *reproducible* artificial morphogenesis that integrates self-assembly and pattern formation under the control of a nonrandom gene regulatory network. The specialized properties of cells (division, adhesion, migration) are determined by the gene expression domains to which they belong, while at the same time these domains further expand and segment into subdomains due to the self-assembly of specialized cells. Through this model, I also promote a new discipline, *embryomorphic engineering* to solve the paradox of "meta-designing" decentralized, autonomous systems.

1 Self-Assembly of Pre-Patterned Components

1.1 From puzzles to self-assembly

In the "jigsaw puzzle" metaphor of self-assembling systems, in particular molecular and biological self-assembly, a "piece" of the puzzle represents an elementary component of the system, such as a molecule or a cell. The "shape" of this piece represents its binding affinities with other components—an electric field in the case of molecules (via ionic or hydrogen bonds) or differential adhesion in the case of cells (via specific membrane proteins). At any instant, the puzzle finds itself in a certain state, which corresponds to a particular spatial arrangement of its pieces. Associating an energy or cost function with states, the "solutions" of the puzzle can then be defined as the energy minima, i.e., those states where all pieces "fit" well together and satisfy each other's constraints.

Naturally, several fundamental aspects also distinguish complex self-assembling systems from jigsaw puzzles (Fig. 1):

(i) **Affinities**: The fit between components is not necessarily all-or-none but approximate or flexible (compare Fig. 1a,d) and may exhibit different degrees of well-formedness, associated with varying energy costs. Thus, the "solutions" of the system need not be strict energy minima but simply low-cost states.

(ii) **Component types**: Components' shapes are far from unique. The system is generally composed of distinct types (molecule species or cell types) shared by a multitude of components that are copies of each other (Fig. 1d-f). This allows for a large number of equivalent states, invariant by permutation of components, and greatly facilitates convergence toward one of the many low-energy solutions.

(iii)**Control**: No centralized control or "visible hand" actually moves the pieces.

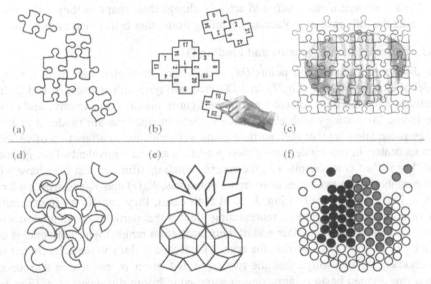

Fig. 1: Differences between jigsaw puzzles and self-assembly. (a)-(c) Jigsaw puzzles are made of uniquely "shaped" pieces, where shape constraints result from specific geometry as in (a), specific markings as in (b) or both geometry and markings as in (c). Compatibility with other pieces is a rare or unique event, and fit between pieces is rigidly all-or-none. Generally, there is only one solution, which requires a long time to find. (d)-(e) By contrast, natural self-assembly (molecular or multicellular) consists of only a few types of identical components—schematized by one type in (d), two in (e), three in (f)—fitting each other tightly in tilings (e) or loosely in aggregates (d) and (f). This multiplicity and flexibility give rise to many possible approximate "solutions" via quick convergence times. Finally, no central process is steering the pieces.

1.2 From molecular self-assembly to multicellular self-assembly

How, then, do self-assembling pieces find their way to their final positions on the basis of purely local interactions and create global order at the system level? At this point, principles of *molecular* self-assembly, on the one hand, and *multicellular* self-assembly, on the other hand, diverge in several important regards:

(iv) **Existence of components**: Molecules generally pre-exist in the solution before they self-assemble. Cells, however, are dynamically created *during* self-assembly by the division of other cells.[1]

(v) **Binding fate**: Molecules initially form a homogeneous mixture (the puzzle box) and bind to each other through stochastic collisions (possibly with help from enzymes, but the original encounter remains stochastic). Cells, however, appear *on the spot*, again by cellular division, in the neigborhood of the cells to which they bind (possibly later changing neighborhood through migration, but this is also a highly nonrandom process).

(vi) **Shape determination**: Possibly folding upon themselves after synthesis, molecules settle on a relatively fixed (passive) geometrical shape and admit only a limited amount of deformation when coming into contact with other molecules. Cells, however, *dynamically* and actively change their shape as they differentiate under the influence of molecular signalling from other cells (such as induction).

1.2.1 Existence of components and binding fate

The distinctions outlined in points (iv) and (v) are illustrated here with a simple model of swarm behavior (Fig. 2). In 2-D space, two types of particles, α and β (respectively dark and light colored spots in the figure), interact via attractive and repulsive forces. By analogy with electric fields, these interactions are modeled as local energy potentials $V(\mathbf{r})$ that each particle emits in its vicinity as a function of vector $\mathbf{r}$ from its center. In this model, interaction potentials are the equivalent of the geometrical "shapes" of components, i.e., the specific binding affinities that they have with their neighbors. The two types of isotropic potentials, $V_\alpha(r)$ and $V_\beta(r)$, with $r = \| \mathbf{r} \|$, used in Fig. 2 are graphed in Fig. 3a,b. In both cases, they contain an impenetrable core of infinite values below r_c, representing the fact that particles are nondeformable discs of radius $r_c/2$. At the other end of their interaction range, their "horizont" is defined by r_0. Beyond that distance, the energy landscape is flat and particles do not see one another. What distinguishes the two types is that an α particle is surrounded with a ring-shaped basin of attraction at some equilibrium distance $r_e < r_0$ (Fig. 3a), while a β particle is not and simply repels other particles that come too close without attracting them (this virtually corresponds to $r_0 < r_e$). The type-α basin is quadratic, simulating the establishment of a spring-like force with resting length r_e as soon as two particles come closer than r_0 (but farther than r_c).

---

[1] Naturally, a new cell does not appear *ex nihilo* but is itself the result of self-assembly at a lower level of molecules synthesized by the mother cell, which draws pre-existing biochemical resources from the extracellular matrix. The present analogy, however, focuses on whole components and whether they are ready to assemble with other components at their own level.

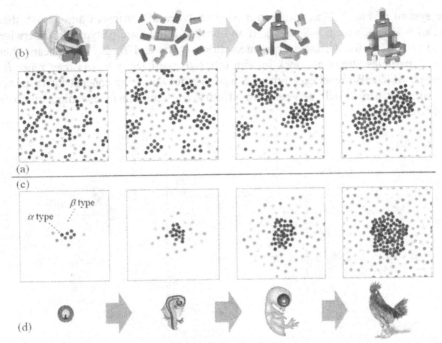

Fig. 2: Constrasting *molecular*-style self-assembly with *multicellular*-style self-assembly by a simple swarm model and metaphorical illustrations. The model contains particles of two types, α (dark color discs) and β (light color discs), that exert an attractive and/or repulsive potential V on their neighbors, with $V = V_\alpha$ for α-α interactions (Fig. 3a with $r_c = 1.6$, $r_e = 3.5$, $r_0 = 5$) and $V = V_\beta$ for α-β and β-β interactions (Fig. 3b with $r_c = 1.6$, $r_0 = 2.5$). Molecular self-assembly (a) relies on a random mix of pre-existing particles that sort out and aggregate through chance encounters. This would be equivalent to shaking a magnetic construction block game (b), in which pieces bear selective geometrical affinities. Multicellular self-assembly (c), however, as in animal development (d), mostly results from *growth* through cell division, not stochastic collisions. New cells are born already pre-positioned and rearrange only locally.

This swarming system is similar to previous models of collective motion in computer graphics [Reynolds 1987] and physics [Vicsek et al. 1995; see, e.g., Grégoire & Chaté 2004], with the difference that in those models particles are self-propelled at constant speed v_0 and only their direction of motion θ_i is updated at every time step. In the present model, the velocity may vary in both norm and direction according to a simplified equation of motion:

$$\lambda \dot{\mathbf{x}}_i = -\sum_j \nabla_i V(\mathbf{x}_j, \mathbf{x}_i) + \eta \qquad (1)$$

where $\mathbf{x}_i$ is the position of particle i, $V(\mathbf{x}_j, \mathbf{x}_i)$ is the potential created by particle j in $\mathbf{x}_i$, and η is white noise. (The above equation can be derived from $m\, d^2\mathbf{x}_i/dt^2 = -\lambda\, d\mathbf{x}_i/dt - \sum_j \nabla_i V(\mathbf{x}_j, \mathbf{x}_i) + \eta$ by neglecting the inertia term $m\mathbf{x}''$ in front of the viscosity term $\lambda\,\mathbf{x}'$, or assuming that particles are quasi-stationary.) Now, in the mixed type system of Fig. 2, $V(\mathbf{x}_j, \mathbf{x}_i)$ depends on *both* types of particles i and j, such that $V(\mathbf{x}_j, \mathbf{x}_i) = V_\alpha(r_{ij})$, with $r_{ij} = \| \mathbf{x}_i - \mathbf{x}_j \|$, if and only if i and j are of type α ; otherwise $V(\mathbf{x}_j, \mathbf{x}_i) = V_\beta(r_{ij})$ for the other type interactions $\alpha\!-\!\beta$ and $\beta\!-\!\beta$. This means that only α particles attract and lock in with other α particles, whereas they repel β particles, which also repel each other. "Shape" is thus a relative concept for α particles, as they can switch between two affinity configurations depending on their neighbor's type.

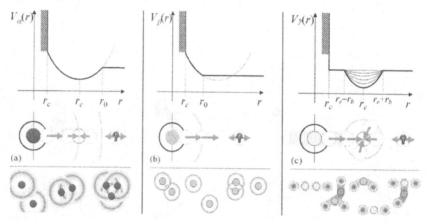

Fig. 3: The "shape" or binding affinities of self-assembling components can be modeled by the emission of local attraction/repulsion potentials V. In each frame: top—graph of V as a function of distance r from the particle's core; middle—2-D view "from above" of a neighbor particle's motion within V; bottom—example of a few particles interacting through V. (a) Isotropic elastic potential used in the $\alpha\!-\!\alpha$ interactions of Fig. 3. (b) Isotropic repulsion used in the $\alpha\!-\!\beta$ and $\beta\!-\!\beta$ interactions of Fig. 3. (c) Anisotropic "polar" attraction potential used in Fig. 4.

As a result, in Fig. 2a, an initial random mix of 110 particles of each type reliably converges toward a lower energy state: α particles collide by chance (random walk due to the stochastic term η), stick to one another and progressively form larger aggregates until a giant component containing all α particles is created in the midst of a sea of β particles. Self-sorting processes such as this one have also been simulated using the Potts formalism [Graner & Glazier 1992], a multivalued Ising model in which a pixel on the grid represents a fragment of a biological cell, and a local region of equal pixel values represents one cell. Instead of the point-wise motion of swarm systems, the Potts model shifts cell boundaries by flipping pixel values according to a stochastic surface energy minimization. Similarly to Eq. (1), Potts surface energy includes differential adhesion as a sum of pairwise interactions between cell types.

Although dissociated and mixed cells can spontaneously sort out again into homogeneous tissues, this phenomenon is seen mainly in artificial experiments whose goal is to demonstrate differences in cell adhesiveness. Cell sorting does not constitute a major natural developmental mechanism—despite common features with cell migration at the level of adhesion proteins. It might intervene locally as a correction mechanism (e.g., to compensate for small fluctuations or errors in cell differentiation) but, in the end, an organism does *not* emerge from a giant swarm of trillions of disaggregated cells that reassemble in parallel. Biological morphogenesis is mostly the product of regulated *growth*, i.e., guided positioning by division and migration, not chance encounters.

For these reasons, Fig. 2a is a more faithful illustration of the *molecular* style of self-assembly (the first half of points (iv) and (v)), while *multicellular*-style self-assembly is better captured by Fig. 2b (the second half of points (iv) and (v)). Here, the system starts out with only a few particles of each type, which later divide into same-type particles according to a certain probability. New particles pop up already pre-positioned near the type that produced them and only briefly rearrange within their local neighborhood. Not addressed in this simple model is cell *migration*, which also plays an important role in animal development. Yet again, migration has little to do with stochastic collisions. After its birth within a given neighborhood, a cell may traverse its environment toward a specific remote location, but it does so only under tight guidance from extracellular signals and cell-to-cell adhesion properties.

1.2.2 Shape determination

Once positioned, biological cells, unlike puzzle pieces or molecules, are also able to modify their individual shape *dynamically* and, consequently, the local geometrical arrangement that they form with their neighbors. This is an important aspect of multicellular self-assembly that was mentioned in point (vi) and will be modeled here by a variant of the previous swarm system. In this variant, illustrated in Fig. 3c and Fig. 4, the isotropic potentials $V_\alpha(r)$ and $V_\beta(r)$ are replaced with an anisotropic or "polar" potential $V_\gamma(\mathbf{r})$. Instead of V_α's ring-shaped basin of attraction at distance r_e from the cell center, the potential landscape V_γ has two localized basins of attraction (quadratic wells of radius r_b) centered around two poles $\mathbf{r}_1 = (\theta_1, r_e)$ and $\mathbf{r}_2 = (\theta_2, r_e)$. For example, in Fig. 3c the values of θ_1 and θ_2 are 0 and π in the case of the horizontal (green) segments. In Fig. 4, the swarm consists of γ particles with "vertical" binding affinities $\theta_1 = \pi/2$ and $\theta_2 = -\pi/2$ (Fig. 4a,c) or variable angles (Fig. 4b,d). To represent the shape of this polar potential, these particles are displayed as short segments of length r_e and thickness r_c, instead of discs. In Fig. 4a-b the particles pre-exist in a mix and have fixed shapes, whereas in Fig. 4c-d they dynamically appear and reshape themselves. Thus, while the contrast between Fig. 4a (colliding) and Fig. 4c (growing) reiterates points (iv) and (v) already illustrated in Fig. 2, here in the case of polar particles, the contrast between Fig. 4b (fixed shapes) and Fig. 4d (dynamical shapes) focuses on point (vi). By dynamically changing their shape after placing themselves into chain formation (through colliding or growing), the particles of Fig. 4d create a specific morphology, *which is otherwise difficult or impossible to attain spontaneously through sheer stochastic encounters* (Fig. 4b). The preshaped particles of Fig. 4b have specific pairs of angles that replicate the final state of Fig. 4d but are unable to coordinate; they only explore suboptimal and unstable states. By contrast, Fig. 4d is analogous to the *invagination* of cell membranes, a common mechanism of animal development most striking during gastrulation, whereby a few cells constrict one of their sides (using filaments of motor proteins) and adopt an elongated "wedge" aspect that draws the entire sheet of neighbor cells towards them.

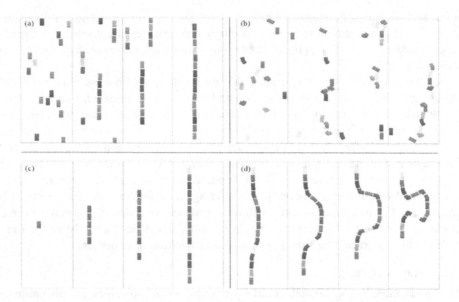

Fig. 4: Constrasting different modes of self-assembly in 1-D. In all frames, particles are of the same type γ, i.e., they interact via the polar potential $V_\gamma(r)$ of Fig. 3c. To remind of their anisotropic affinities, particles are drawn as small rectangles instead of discs. (a) Colliding self-assembly: 15 particles with vertical poles $(\theta_1, \theta_2) = (\pi/2, -\pi/2)$ quickly snap into place, forming a straight chain. (c) Growing self-assembly: as in Fig. 2c, the same string can be formed by dividing particles. (d) Reshaping self-assembly: each particle of (c) now dynamically bends its shape in specific ways (see Fig. 5c), making the string invaginate. (b) Preshaped self-assembly: the invagination *cannot* be reproduced by giving fixed angles to the particles in advance (the same angles that appear at the end of (d)) and letting them randomly collide.

In summary, biological self-assembly at the cellular level relies on principles that greatly facilitate and accelerate morphogenesis. When designing self-organized artificial systems in future, letting components dynamically create and reshape themselves "on the spot," as cells do, would be a far more efficient approach than letting them haphazardly try to match each other's pre-existing constraints, like molecules in a solution. (Obviously, a major technical difficulty will be to implement and control the self-replication of artificial components made of electric, chemical or even biological materials.) In any case, to transition from *stochastic* (molecular-style) self-assembly to *programmable* (multicellular-style) self-assembly, components must be able to modify their behavior (divide, differentiate, migrate) dynamically through *communication*. Cells do not just snap into place; they send molecular signals to each other, forming patterns of differentiation at the same time that they are self-assembling.

2 Pattern Formation in Pre-Assembled Media

Since Turing's 1952 seminal model of the spontaneous symmetry breaking and appearance of regular structures in biological organisms, the concept of *morphogenesis* has largely, but somewhat inaccurately, become synonymous with *pattern formation*. Morphogenesis originally referred to the biological development of the organs and structures of an organism during embryogenesis and, by extension of its etymology, any "generation of form" at various scales in other types of complex systems, such as physical (geomorphogenesis) or social (urban morphogenesis). Pattern formation, in contrast, generally refers to the emergence of statistically regular motifs in quasi-continuous and initially homogeneous 2-D or 3-D media. To be sure, both phenomena involve the decentralized self-organization of a myriad of elements and produce contrast where there was uniformity, yet they do not emphasize the same aspect of emerging order. The latter looks at shimmering landscapes of *activity* on a more or less fixed backdrop, while the former emphasizes the creation of intricate *architectures* and structures. Using an artistic metaphor, it could be said that pattern formation "paints" a pre-existing space, while morphogenesis "sculpts" its own space.

There is a huge diversity of pattern formation behaviors across many scales and substrates (e.g., fluid, electromagnetic, mechanical, chemical, biochemical), from which a few broad classes of mechanisms and models have been identified (e.g., convection cells, reaction-diffusion, activator-inhibitors, synchronization of oscillators). The observed patterns can be static, steady-state or dynamically changing (e.g., traveling waves) and organize themselves into patches or domains that also fall into a few classical geometrical families (e.g., spots, stripes, spirals, branches). Morever, the pattern formation processes typically studied are for the most part inherently *stochastic*, at both the microscopic level of elements and the macroscopic level of the distribution of patterns. Continuing the tradition initiated by Turing, most models have been focusing on systems that rely on instabilities and *amplification of fluctuations* to transition toward order and form patterns. Because of their randomness, and without carefully set boundary conditions (possibly themselves the product of morphogenesis; see below), the outcome of those processes is generally unpredictable in the number and position of the emerging mesoscopic domains (spots, stripes, convection cells). At the same time, the whole formation on the macroscopic level is fairly regular or even periodic, at least piecewise, since it essentially consists of repeated motifs. It displays statistical uniformity similar to textures.

In biological development, by contrast, the mesoscopic elements (organs, limbs, parts, tissues, etc.) are always very reliably positioned—unlike random spots and stripes—and display complex and heterogeneous morphologies—unlike uniform textures. Although the well-known colorful animal coats, such as seashell, zebrafish or leopard, have been (debatably) assumed to arise from morphogen-based reaction-diffusion pattern formation, they make up only a minor part of the whole organism. The unique characteristic of biological *morphogenesis*, absent from simpler physical-chemical pattern formation, is that each one of its self-organizing elements, the cell, contains a rich source of information stored in the DNA, which endows it with a vast repertoire of highly nontrivial behaviors. Even admitting that DNA is less than a "program," in the sense that it does not control the cell deterministically along a linear flow of execution, it is nonetheless, at the very least, a *repository of stimuli-response rules*, vastly superior in quantity of functional information to any physical or chemical element involved in one of the habitual pattern formation dynamics.

Embryogenesis, therefore, combines both pattern formation and morphogenetic self-assembly in a tightly integrated loop: It creates shape from patterning (Fig. 5) and patterns from shaping. Structures are "sculpted" from the assembly of elements that have been prompted to do so by the "painting" of their genetic identity. Conversely, newly formed shapes are able to support, and trigger, new domains of genetic expression. DNA is "consulted" at every step of this exchange, in every cell, to produce the proteins that will guide the cell's highly specific biomechanic behavior (shaping) and signalling behavior (patterning) at a given time in a given location, depending on the signals received from the neighborhood. A schematic illustration of the "shape from patterning" process is shown in Fig. 5.

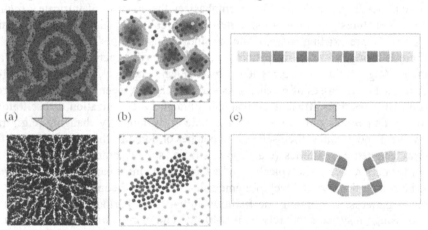

Fig. 5: "Shape from patterning" examples: deriving morphogenetic self-assembly (bottom frames) from pattern formation (top frames). (a) Amoebae in the slime mold *Dictyostelium* first generate waves of chemical signalling, modeled as a lattice of coupled oscillators (top). After a while, the lattice breaks up as cells follow the concentration gradient toward wave centers and aggregate (bottom). (snapshots from T. Schmickl's online simulations at http://zool33.uni-graz.at/schmickl) (b) Augmented view of the swarm of Fig. 2a, where the α particles are assumed to have differentiated from a prepattern of chemical concentration before assembling as above. (c) Augmented view of the chain of Fig. 4d, in which the bending angle of each cell is also determined by a prepattern of identity. (The type of pattern formation in (b) and (c), whether Turing-like, genetically regulated or a combination of the two, is not specified here.)

3 Integrating Self-Assembly and Pattern Formation Under Genetic Regulation

The model of artificial embryogenesis that I have recently proposed [Doursat 2006, 2007] is an original attempt to integrate the three fundamental ingredients discussed above: (i) self-assembly (SA) and (ii) pattern formation (PF), triggering each other in a feedback loop under the tight control of (iii) nonrandom genetic instructions (GI)—here, a gene regulatory network—stored in each cell of the system. Previous theoretical models of biological development or bio-inspired artificial life systems have seldom included all three mechanisms. The evo-devo works of Hogeweg [2000], Salazar-Ciudad & Jernvall [2002] or, with lesser morphogenetic abilities, the *Cellerator* system [Shapiro et al. 2003] and Nagpal's origami [2002] are a few notable achievements. Other interesting studies have explored the combination of two out of three ingredients: SA and PF, no GI—self-assembly based on cell adhesion and signalling pattern formation, but using only predefined cell types without internal genetic variables [e.g., Marée & Hogeweg 2001]; PF and GI, no SA—nontrivial pattern formation from instruction-driven intercellular signalling, but on a fixed lattice without self-assembling motion [e.g., von Dassow et al. 2000, Coore 1999]; SA and GI, no PF—heterogeneous swarms of genetically programmed, self-assembling particles, but in empty space without mutual differentiation signals [e.g., Sayama 2007].

The present model has been explained in detail elsewhere [Doursat 2006, 2007]. It is summarized here, highlighting the interplay between pattern formation and self-assembly, as illustrated in Fig. 6.

3.1 Gene-regulated pattern formation

A virtual embryo is a swarm of cells, where each cell contains a *gene regulatory network* (GRN) coding for its signalling and mechanic activity. Through intercellular coupling between neighboring GRNs, the embryo becomes patterned into *identity domains* of differentiated gene expression, creating a "hidden geography" revealed by *in situ* hybridization (Fig. 6c,e). Essentially, logical combinations of regulatory switches (OR, AND) translate into geometric combinations of precursor patterns into new patterns (union, intersection). Developmental genes are roughly organized in tiers, or "generations." Earlier genes map the way for later genes and gene expression propagates in a cascade. This principle has been beautifully demonstrated in the *Drosophila* embryo. The intersection of various striping patterns along its three main axes gives rise to smaller domains such as the organ primordia and "imaginal discs," which are groups of cells marking the location and identity of the fly's future appendages (legs, wings, antennae). Going back in time, the whole process begins with concentration gradients of maternal proteins diffusing across the initial cluster of cells. These gradients are the functional equivalent of a coordinate system.

3.2 Biomechanic self-assembly

In parallel to genetic patterning, the embryo continues to grow and undergo extensive reshaping as cells divide and proliferate. Previous identity domains expand and deform while becoming partitioned into new and finer identity domains. Three main biomechanic principles responsible for these morphogenetic changes are integrated into the model as schematic rules: (a) differential *adhesion* as elastic forces between cells, (b) inhomogeneous cell *division* as internal probability rates, and (c) tropic cell *migration* as internal velocity vectors. In parallel to chemical coupling between their GRNs, neighboring cells are connected by abstract mechanical edges between their nuclei, established through Delaunay triangulation—the cell shapes being the complementary Voronoi domains. Similarly to the $V_\alpha(r)$ potential of Fig. 3a, cell-to-cell edges are modeled as small springs, so that each cell tries to set the distance with its neighbors' nuclei to r_e. Biological cells also stick to one another by means of adhesion proteins that cover their membrane. The great diversity of surface adhesion proteins gives them the ability to selectively recognize one another by modulating their degree of "stickiness." In the elastic force model, differential adhesion on edge $i \leftrightarrow j$ between cells i and j is modeled by resting lengths l_{ij} spring coefficients k_{ij} that can vary from edge to edge. Different proliferation frequencies ρ_i also create deformation in the embryo, as compartments expand faster than others. Migration, represented by additional time- and space-varying vectors v_i, is not shown in this article.

3.3 Pattern-regulated self-assembly, assembly-triggered patterns

Closing the loop, the complete model *establishes a functional dependency between cell identities and mechanical cell behaviors* (Fig. 6). The self-assembly rules of section 3.2 are linked to the self-patterning process of section 3.1 by making mechanical parameters ρ_i, l_{ij} and k_{ij} depend on the current state of the genetic expression of i and j, i.e., the identity (colored) domains to which cells belong. See Fig. 6 for details.

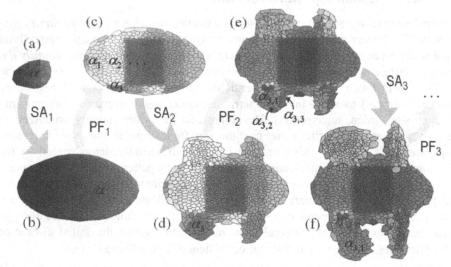

Fig. 6: Integrating self-assembly (SA) and pattern formation (PF) under genetic regulation. (a) Starting from a small clump, cells proliferate at a uniform division rate to reach about 800. All cells are of same type α with an anisotropic 2:1 ratio along x (ellipsoid version of elastic potential V_α, Fig. 3a). (b) As in Fig. 2b, spontaneous rearrangements give the embryo a convex shape, here oval. During expansion, protein gradients (x-gradient in purple) spread across the unique domain. (c) The varying concentration levels are then read out by each cell and input into the first stages of their GRNs (not shown), producing in output different values of gene expression, i.e., defining new *cell types* α_1, α_2 and so on. (d) These types, in turn, determine the new division rates $\rho(\alpha)$ in each domain and adhesion coefficients $k(\alpha, \alpha')$ between domains. For example, α_3 and three other types start proliferating at a faster rate than the rest of the embryo, while they lose adhesion with neighboring domains, thereby creating limb-like bulges. (e)-(f) The same alternation of PF-induced differentiation and heterogenous-type SA

continues at a finer scale of detail with respect to the overall expanding organism.

4 Toward Evolutionary Meta-Design

Exploding growth in hardware components, software modules and network users will force us to find an alternative to rigidly designing and controlling computational systems in every detail. Instead, future progress in information and communication technologies will depend on our ability to *meta-design* mechanisms allowing those systems to self-assemble, self-regulate and evolve. Nature offers a great variety of decentralized, autonomous systems, most conspicuously biological organisms. Deemed "complex," these systems might in fact be less costly, more efficient or even "simpler" than human-designed and centrally controlled contrivances. Complex systems are characterized by the self-organization of a great number of small, repeated elements into large-scale, adaptive patterns, where each element may itself obey the dynamics of an inner network of smaller entities at a finer scale (microprogram). The new engineering challenge is thus to "guide" this self-organization, i.e., to prepare the conditions and mechanisms favorable to nonrandom, heterogeneous and reproducible morphogenesis (macro-program). At the same time, it is also to allow the parameters of this process evolve in order to freely generate innovative designs. Finding efficient systems will require matching loose selection criteria with productive variation mechanisms. The first point concerns the openness of the designers to "surprising" outcomes; the second point concerns the intrinsic ability of complex systems to create a "solution-rich" space [Minai et al. 2006] by combinatorial tinkering on highly redundant parts. Embryogenesis, the development of an entire organism from a single cell, provides the most striking example of self-organization guided by evolvable genetic information.

This work describes an original model of bio-insipred, artificial *embryomorphic* system growth, integrating pattern formation and self-assembly under non-random genetic regulation. A virtual organism is represented by a mass of cells that proliferate, migrate and self-pattern into differentiated domains. Each cell contains an internal gene regulatory network and acquires a specific gene expression identity by interaction with neighboring cells. Differentiated cell types trigger different cell behaviors, which in turn induce new identities. The organism's final architecture depends on the detailed interplay between the various rates of cell division and movement, propagation of genetic expression and positional information. Ultimately, on this score of "theme and variations" (developmental laws and parameters), evolution will be the player.

Based on these first results, I propose a new discipline, *embryomorphic engineering* as a "fine-grain" approach to systems design, based on swarms of relatively simple, cloned elements. It emphasizes the need for hyper-distributed architectures and self-organized development as prerequisites for evolutionary innovation. In possible future hardware applications, nano-units containing the same instructions could self-organize without the need for reliability or precise arrangement as in traditional VLSI [Coore 1999, Nagpal 2002]. In software or network applications (servers, security, etc.), a swarm of small-footprint software agents could diversify and self-deploy to achieve a desired level of functionality.

Bibliography

[1] Coore, D., 1999, *Botanical Computing: A Developmental Approach to Generating Interconnect Topologies on an Amorphous Computer*, Ph.D. thesis, MIT Department of Electrical Engineering and Computer Science.

[2] Doursat, R., 2006, The growing canvas of biological development: Multiscale pattern generation on an expanding lattice of gene regulatory networks, *InterJournal: Complex Systems*, **1809**.

[3] Doursat, R., 2007, Organically grown architectures: Creating decentralized, autonomous systems by embryomorphic engineering, In *Organic Computing*, R. P. Würtz, ed., Springer-Verlag, to appear.

[4] Graner, F., & Glazier, J. A., 1992, Simulation of biological cell sorting using a 2-D extended Potts model, *Physical Review Letters*, **69**(13): 2013–2016.

[5] Grégoire, G., & Chaté, H., 2004, Onset of collective and cohesive motion, *Physical Review Letters*, **92**: 025702.

[6] Hogeweg, P., 2000, Evolving mechanisms of morphogenesis: On the interplay between differential adhesion and cell differentiation, *Journal of Theoretical Biology*, **203**: 317–333.

[7] Marée, A. F. M., & Hogeweg, P., 2001, How amoeboids self-organize into a fruiting body: Multicellular coordination in *Dictyostelium discoideum*, *PNAS*, **98**(7): 3879–3883.

[8] Minai, A. A., Braha, D., & Bar-Yam, Y., 2006, Complex engineered systems, in *Complex Engineered Systems: Science Meets Technology*, D. Braha, Y. Bar-Yam and A. A. Minai, eds., Springer-Verlag.

[9] Nagpal, R., 2002, Programmable self-assembly using biologically-inspired multi-agent control, *1st Int Conf on Autonomous Agents*, Bologna, July 15-19.

[10] Reynolds, C. W., 1987, Flocks, herds and schools: A distributed behavioral model. *Computer Graphics*, **21**(4): 25–34.

[11] Salazar-Ciudad, I., & Jernvall, J., 2002, A gene network model accounting for development and evolution of mammalian teeth, *PNAS*, **99**(12), 8116–8120.

[12] Sayama, H., 2007, Decentralized control and interactive design methods for large-scale heterogeneous self-organizing swarms, *Advances in Artificial Life: Proceedings of the 9th European Conference on Artificial Life (ECAL)*.

[13] Shapiro, B. E., Levchenko, A., Meyerowitz, E. M., Wold, B. J., & Mjolsness, E. D., 2003, Cellerator: Extending a computer algebra system to include biochemical arrows for signal transduction simulations, *Bioinf*, **19**(5): 677–678.

[14] Vicsek, T., Czirók, A., Ben-Jacob, E., Cohen, I., & Shochet, O., 1995, Novel type of phase transition in a system of self-driven particles, *Physical Review*

Letters, **75**: 1226–1229.

[15] von Dassow, G., Meir, E., Munro, E. M., & Odell, G. M., 2000, The segment polarity network is a robust developmental module. *Nature* **406**: 188–192.

Chapter 7

The Necessity of Conceptual Skill Enhancement to Address Philosophical Challenges of New Science : Background and Implications

Robert A. Este
Institute for Biocomplexity and Informatics
The University of Calgary
reste@ucalgary.ca

Sensory extension and increasingly powerful computation underpinning emerging new science provides us with daunting conceptual challenges. It is here argued that because they are primarily philosophical in nature, high-level conceptual challenges are unique and different from those that help us refine outputs and achievements based on solving our technical and political problems. This paper explores the practical implications of such unique conceptual challenges and discusses the need for enhancement of conceptual skills. Such conceptual enhancement may provide us with a good opportunity to best respond to and benefit from these challenges.

1 Introduction

The use of information to achieve knowledge, derived from data originating with many sources, is essential to successful enterprise in any human effort. Indeed, we note that the derivation and use of knowledge is key to human advance in any field. This appears to be obvious, but the central argument provided in this paper is that the obviousness of this phenomenon that leads us through our processes of scientific discovery and advance may mask the emergence of unique conceptual challenges to our conceptual skill capacity defined by their philosophical nature, where such challenges are not at all obvious.

2 Background

To accomplish the derivation of scientific knowledge, we engage in two essential activities of science: [i] we continue to extend our senses to gather increasingly detailed, voluminous and complex data sets having to do with the very large and the very small, and the very complex and the very dynamic; and [ii], we apply many tools of analysis and synthesis to those data, and in the modern era we continue to develop and employ increasingly complex, networked, and very powerful advanced perceptual and computational tools for this purpose, especially those having to do with simulation, to assist us with and help us continue to develop and explore how to enhance theory and knowledge yields from these fundamental tasks. We create, develop, enhance and strengthen our perceptual and computational tools to the extent that, today, we are capable of deriving very useful and sometimes unexpected and exceptionally valuable information about extremely complex systems comprised of vast numbers of interacting components and agents [1]. This is important exploratory work that we could not accomplish, or even dream of accomplishing, only a relatively short time ago [2].

Prior to the advent of the current era where both highly advanced sensing devices coupled with very powerful computational machines continue to be developed, not much could be known about either the working intricacies of high dimensional complex systems, or of the extent and emergent nature of highly complex interactions among large numbers of high dimensional complex system variables and components. This meant that high levels of functional detail about complex systems might have been postulated, or how such systems behave might have been contemplated, but such detail or postulation was either inaccessible or undeveloped – with the exception, perhaps, of those blessed with a capacity for genius who, based on available early evidence, might be able to work out plausible hypotheses and develop new and useful mathematical models more effectively than others [3]; but even geniuses would not have privileged access to tools of extended perception and advanced computational capacities in advance of everyone else. Such characteristics and features of complex systems were hidden behind the limits to human perception, were postulated hypothetically by those who had the capacity to push to the edges of the hypothesis development envelope, and awaited discovery beyond a computational ceiling that could begin to be penetrated only by advanced mathematical modeling of plausible systems – but again, in relative terms, this ceiling was kept low by early generation computational capacities. Knowledge about what actually comprised such complex systems and their environments, as well as their behaviours – or even knowledge [not speculation] that such systems themselves existed – could not be accessed or perhaps not even surmised except in very special circumstances of hypothesis construction, and certainly not beyond rational conceptual limits which would relegate such thinking to the realm of fantasy.

Such complex systems, when rationally contemplated, could be dealt with only at relatively high levels of plausible abstraction, or, when available, in terms of their known components where, for example, limited principles of structure, logic and function might be explored mathematically and fundamental aspects of emergent theory might consequently be developed [4]. This meant that complexity theory – that is, any adequate theory building having to do with high dimensional, extremely complex systems with large numbers of variables – would initially be difficult in any scientifically meaningful fashion and could not be verified through experiment or simulation except in the most general terms, and therefore, almost exclusively, were known only in principle. For example, features and behaviours of complex systems such as ecologies, cortical columns, weather systems, genetic regulatory networks, quantum chemical interactions, dynamical activities among subatomic particles, energy and work networks of a cell, the Internet, or hitherto unknown dynamical astronomical objects could, in essence, be "sketched out" conceptually and perhaps mathematically modeled in some useful but limited fashion – and, all had in common the feature of residing in and perhaps even defining the archetypical "black box" [although of course as we well know, it is a very important step forward in knowledge generation when a black box is recognized where before none was perceived or even postulated!] [5].

Keeping these thoughts having to do with the advance of science as a backdrop to this paper, we note that, in the current era, based on the development and use of tools that extend our perceptual capabililty and provide advanced computation that together do such a good job at enhancing our productivity, creativity and problem solving capacities, we continue to explore, discover, invent and develop at a rapid pace. In other words, science continues to rapidly move forward on increasingly broad and comprehensive fronts and its valuable outputs proliferate. It is no surprise to note that part of this is taking place with regard to the very perceptual and computational tools we employ. Although I do not here address any details of the technical enhancement of computation and perception, a fundamental question with two related parts underpins what is addressed here: [i] what frameworks will allow us to best understand the components that allow us to make scientific advances; and [ii], if it is the case that we continue to develop our scientific capacities and knowledge, what is the extent and nature of challenges that present themselves as this development takes place, and how can we make best use of them? Let us explore where these formative questions lead us.

3 The Challenge

I have argued elsewhere [6] that conceptual and philosophical challenges in the quantum mechanical and holonic enterprise fields illustrate a serious problem having to do with scientific advance; these two realms have a striking isomorphism with the fundamental two-part question outlined above. That is, given unfolding new knowledge about complex systems in many realms of science such as those mentioned earlier, and the improving tools we use to explore this diverse realm, we are faced with the challenge of developing a comprehensive conceptual foundation that would successfully account for this emerging "new science", a foundation that goes beyond what I am here describing as the type of conceptual work necessary for spectacular technical scientific advance such as what was accomplished by the Manhattan Project and American moon landing teams, for example [7]. By this I mean that solving a vast array of technical, mathematical and physics problems having to do with complex systems as identified above, and coming up with novel solutions to those problems, does indeed take advanced conceptual skills of a particular kind rooted in and colored by the very technologies we develop [e.g., we need to think effectively about mathematical models and the engineering of new tool development to permit us to work in the particular complex systems realm of interest] – and clearly, we have to be very smart and well-trained to accomplish such tasks. However, we must at the same time be very cautious about assuming that our extensive technical [and even political] work that results in accumulations of good prescriptive knowledge [8] in the multiplicity of complex systems realms, some of which are identified above, will automatically result in some form of a solid conceptual foundation that permits us to stand confidently on a new epistemology of complex systems which in turn would be a specific example of the general case of the emergence of new science [see, for example, Kuhn (9), Nickles (10) and Suppe (11)].

The point here is that, from an epistemological perspective, there is a potentially serious danger of assuming that new propositional knowledge [8] derived from what is learned from vast amounts of new accumulated prescriptive knowledge that has originated from extended sense perception and computational capacity is the equivalent of an adequate conceptual foundation capable of supporting new epistemologies of new science [12]. Assuming that this is the case would amount to "masking" the authentic conceptual challenge of emerging new sciences of complexity, for example, which is what I have suggested elsewhere may also be taking place with regard to quantum physics and the holonic enterprise: in other words, new propositional knowledge about complex systems derived from extensive investigation and work in specific fields of inquiry can potentially be mistaken for the conceptual adequacy necessary to effectively deal with and account for new epistemologies of new science. This may be a state of misapprehension into which many contemporary scientists might easily fall, and I am among those ranks; it is, I think, natural to assume that if truly difficult scientific problems have been "cracked" and new emerging science has indeed been developed and advanced, especially problems having to do with complex systems on "the leading edge" that require advanced tools of perception and computation, that such advance is itself necessary and sufficient evidence to qualify as well-understood progress in the essential philosophical foundation of science [13]. However, I do not think that this is the case; in fact, I believe that this is a philosophical error that may be more common than we would like to think, and we are compelled to guard against falling victim to it. As I have stated elsewhere [6], "This type of conceptual slippage is of great concern in the emergence of new science for it speaks to the potential for making important conceptual errors that are generally not perceived, and this has great significance in the context of how we then actualize new science in our organizations and diffuse new science throughout our cultures." In other words, if we do good work in a field of emerging science such as the complexity sciences, and if we become a part of a growing community of investigation and enterprise based on this field of emerging science and are rewarded in these pursuits, because of the above-mentioned conceptual slippage we may lose sight of the philosophical foundation of that science and therefore arrive at and commit to non-optimal decisions about how further investigation into that emerging field should be funded, supported, criticized, and, in the end, how that emerging field should be understood. It may be the case, in other words, that without careful consideration of the philosophical foundation of new emerging science, we may not think well enough about how to make good and effective use of that emerging science. This possibility has many scientific, economic and societal implications.

4 Conclusion

The challenges of scientific advance include how well we develop and make use of new tools of analysis and synthesis that permit us to explore new fields and derive new data that we can then reduce to new information about our objects of interest and investigation, whether they be cancer stem cells or micro-loan economies, plasma furnaces or climate models, stellar atmospheres or human consciousness, predator-prey relations or communication network dynamics. From these investigations we derive everything from methods to achieve economic advance and sustainability to new scientific paradigms. But the challenges of scientific advance also include our collective conceptual capacity to understand and shape an emerging epistemology of new science. I do not believe we can have one without the other; nor do I believe we can afford to be blind to the relationship between the two. It is our higher-level conceptual capacity to deal with issues of the epistemology of science, and especially our shared awareness of this capacity, that I suggest we need to refine and build in order to not slip into a state where we are compelled to be satisfied with the default outcomes of scientific advance, founded on fine technically-focused accomplishment coupled with the error of mistaking such accomplishment for philosophical clarity. To assume that this argument is unimportant, or that it is not necessary to think about and then make good use of enhanced understanding of the philosophy of science regardless of our fields of specialization, or that the philosophy of science is essentially a passive enterprise and will take care of itself without much thought are, I believe, components of a serious and in fact costly error. Good science proceeds and advances not only with creative investigation coupled with innovative and productive solutions to perplexing problems in many fields [as well as new interdisciplinary fields], it proceeds with a robust foundation of philosophical clarity about our higher-level conceptual skills [14], and, especially, clarity about how to use those skills to our best advantage. Placing the challenge of epistemological awareness in a more prominent place in our consideration of and actions taken to support scientific advance will, I believe, help us think with enhanced high-level conceptual clarity about our sciences in general and the full extent of scientific, economic and societal advances, outcomes and benefits we aim to achieve. Commencing with complexity sciences in this way may be a very good place to start.

Bibliography

[1] Este, Robert A., "The Significance of Long-term Science Policy Considerations in the Development of Systems Biology," conference presentation, IBI / IBM Systems Biology Workshop, IBM Industry Solutions Lab, Hawthorne, NY, USA, [September 18-20, 2006]

[2] Messina, Paul. "Global Trends in Technology and the Cyber Infrastructure Project in The United States," discussion support document, Argonne National Laboratory [February 24, 2004]
<http://www.naregi.org/papers/data/NAREGIsymp2004Messina.pdf>, retrieved July 18, 2006

[3] Hadamard, Jaques. The Psychology of Invention in the Mathematical Field. New York : Dover Publications [1945]

[4] Casti, John. Complexification. New York : Harper Collins [1994]

[5] Magnani, Lorenzo, and Nancy J. Nersessian [Eds]. Model-Based Reasoning : Science, technology, values. New York : Kluwer Academic [2002]

[6] Este, Robert A., "Enhancing 'New Science' Outcomes of the Policy Process: Some Philosophical Problems," Proceedings of the Fourth International Workshop on Computational Systems Biology / WCSB 2006, Tampere, Finland [June 12-13, 2006]

[7] Gleick, James. Genius : The life and science of Richard Feynman. New York : Vintage [1993]

[8] Mokyr, Joel. The Gifts of Athena : Historical origins of the knowledge economy. Princeton and Oxford : Princeton University Press [2002]

[9] Kuhn, Thomas. The Structure of Scientific Revolutions [3rd Edition]. Chicago : The University of Chicago Press [1966]

[10] Nickles, Thomas [Ed.] Thomas Kuhn. Cambridge, UK : Cambridge University Press [2003]

[11] Suppe, Frederick [Ed.] The Structure of Scientific Theories. Chicago : The University of Illinois [1977]

[12] Cetina, Karin Knorr. Epistemic Cultures : How the sciences make knowledge. Cambridge, MA : Harvard University Press [1999]

[13] Conant, James, and John Haugeland [Eds] The Road Since Structure : Thomas S. Kuhn. Chicago : The University of Chicago Press [2000]

[14] Gärdenfors, Peter. Conceptual Geometry : The geometry of thought. Cambridge, MA : MIT Press [2000]

Chapter 8

Evolutionary Perspective on Collective Decision Making

Dene Farrell[1], **Hiroki Sayama**[1], **Shelley D. Dionne**[2],
Francis J. Yammarino[2], **David Sloan Wilson**[3]
[1] Department of Bioengineering
[2] School of Management / Center for Leadership Studies
[3] Departments of Biological Sciences and Anthropology
Binghamton University, State University of New York

Team decision making dynamics are investigated from a novel perspective by shifting agency from decision makers to representations of potential solutions. We provide a new way to navigate social dynamics of collective decision making by interpreting decision makers as constituents of an evolutionary environment of an ecology of evolving solutions. We demonstrate distinct patterns of evolution with respect to three forms of variation: (1) Results with random variations in utility functions of individuals indicate that groups demonstrating minimal internal variation produce higher true utility values of group solutions and display better convergence; (2) analysis of variations in behavioral patterns within a group shows that a proper balance between selective and creative evolutionary forces is crucial to producing adaptive solutions; and (3) biased variations of the utility functions diminish the range of variation for potential solution utility, leaving only the differential of convergence performance static. We generally find that group cohesion (low random variation within a group) and composition (appropriate variation of behavioral patterns within a group) are necessary for a successful navigation of the solution space, but performance in both cases is susceptible to group level biases.

1 Introduction

Collective decision making is becoming more central and indispensable in human society as modern problems increasingly involve interactivity and inseparability within large scale tasks [4]. In high-tech product and software development, for example, the amount of workers participating in a design project can be in the order of thousands as a result of a product's complexity exceeding an individual's capacity, which almost inevitably results in suboptimal outcomes [5]. More recently, online collective decision making among large populations of anonymous participants via computer mediated networks has been implemented for product rating and common knowledge base formation. Both individual behavior and the organizational structure greatly influence decision processes. The complexity of the processes is more manifested when constituents of groups are heterogeneous with regard to both their world views and behavioral propensities. Collective human decision making in such conditions is poorly understood, being one of the most significant challenges in the social sciences.

The leadership, psychology and organizational behavior/management disciplines have examined collective dynamics using both experimental and applied studies. They generally emphasize linear statistical relationships of team and individual level variables [4], without accounting for nonlinear processes, high-dimensional problem space and non-trivial social structure. Complex, nonlinear problem space has been considered in dynamical modeling studies [5], in which interdependence of aspects of problem are considered, but not nontrivial social interactions.

Here we investigate collective decision making dynamics from a novel perspective by shifting the focus of agency from group members to potential solutions being discussed. The decision making processes are described using concepts in evolutionary theory, where evolution acts on a population of potential solutions through mechanisms of selection and variation as effected by human discussants. Group members thus serve both as an evolutionary environment and as implements of evolutionary action on a population of solutions. Within this context, several evolutionary operators can be mapped to human behaviors. Examples include replication (advocacy of an existing idea), subtractive selection (criticism against an existing idea), mutation (revision of an existing idea) and recombination (creation of a new idea by mixing existing ideas).

2 Model

2.1 Groups

We apply evolutionary framework to model simple group decision making processes within a small-sized, well-connected social network structure. We conduct multiple levels of analysis [1,7] on how homogeneities or heterogeneities of world views/goals among the participating agents, as well as group-level behavioral patterns and biases, affect the decision making dynamics and the final outcomes [2].

When group members are heterogeneous in world views, differences between individual utility functions play a crucial role in determining the group dynamics; the relevant level of analysis is within groups. Each member acts as "group parts" [1] to achieve individual objectives. Conflicts of interest make the problem space more complex than that of groups consisting of homogeneous, world perspectives. On such complex landscapes there is more possibility for populations to become stuck at local optima, detrimenting the overall adaptiveness. Contingently, the importance of variation relates to escaping from the local optima in order to reach better solutions.

If group members are homogeneous in their world view, they behave as "group wholes" [1]; the relevant level of analysis is between groups. The population of solutions evolves to adapt to a single utility function shared by all the group members, so the problem space would be simpler than with heterogeneous groups. With little conflicts of interest, selection is relatively important to adaptiveness as speeding up the convergence of discussion. Variation still holds importance, especially with complex nonlinear problems.

Our model assumes that groups are initiated with a list of randomly generated ideas, whereupon they begin to perform a set of actions on the existing population of solutions repeatedly for a fixed number of iterations. Individuals always act in the same order and groups always demonstrate a full rotation. The number of actions on the population of solutions is a product of the number of group members, N, and the number of iterations, t.

In the population, there may be multiple copies of the same type of solution, which represents the relative popularity among group members. Each action is performed on a single copy of solution, not on an equivalence class of all solution replicates.

2.2 Utility Functions

Groups are situated in an M-dimensional binary problem space, with 2^M possible solutions. For a simulation, every solution has a utility value specified by a master utility function U that is unavailable to group members. Individuals perceive solution utility values based on their own utility functions U_j constructed by adding noise to U. We develop a semi-continuous assignment of utility values in the problem space in the following way. First, s representative solutions $S = \{ v_i \}$ ($i = 1 \ldots s$) are generated as random bit strings, where each v_i represents one solution made of M bits. One solution is assigned the maximum fitness value, 1, and another, the minimum fitness value, 0. The remaining $s - 2$ solutions are assigned a random real value between 0 and 1, ensuring that the entire range of utility values is from 0 to 1, for the sake of comparisons between simulation results.

The utility values of all possible solutions in the domain of the master utility function are defined by interpolation using the utility values of representative solutions in S. We use the Hamming distance as a measure of dissimilarity between two bit strings. With this measure, the utility value of each possible solution not present in S is calculated as a weighted average of the utility values of the representative solutions calculated as follows:

$$U(\upsilon) = \frac{\sum_{i=1}^{s} U(\upsilon_i) \cdot D(\upsilon_i, \upsilon)^{-2}}{\sum_{i=1}^{s} D(\upsilon_i, \upsilon)^{-2}} \qquad (1)$$

where $\upsilon \notin S$ is the solution in question, $U(\upsilon_i)$ is the utility of a representative solution υ_i in S, and $D(\upsilon_i, \upsilon)$ is the Hamming distance between υ_i and υ.

Each individual in a group will unconsciously have a different set of utility values for the possible solutions of the problem. Individual utility functions $U_j(\upsilon)$ ($j = 1...N$) are generated by adding random noise to the master utility function so that:

$$U_j(\upsilon) \in [\max(U(\upsilon) - v, 0), \min(U(\upsilon) + v, 1)] \qquad (2)$$

for all υ, where v is the parameter that determines the range of noise. Individuals do not access global maximum/minimum utility values, though they can retrieve a utility value from the function when a specific solution is given.

In addition to individual deviations from a common master utility function, we investigate the effect of common deviations from the "true" utility function, or group level biases. For simulating group level bias we introduce a new step in the generation of individual utility functions, in which the master utility function $U(\upsilon)$ differs from the original true utility function, $U_T(\upsilon)$. Specifically, a bias β is imposed on the true utility function both by flipping bits with probability 0.25 β per bit and adding a random number ranged $[-\beta, \beta]$ to utility values. Solution sets are renormalized to the range [0, 1]. The master utility function is generated from the biased representative solution set. Subsequent methods follow as described above. Bias represents fidelity of information at the group level, where $\beta = 0$ denotes perfect information, and complete randomization is asymptotically approached as bias increases.

2.3 Evolutionary Operators

We identify six evolutionary operators representing individual behaviors reflecting selection or variation. Some operators use a preferential search algorithm to stochastically search the solution population, where r_p solutions are randomly selected and ranked according to their perceived utility values, and then the best or worst solution is selected depending on the nature of the operator being executed.

Replication. Replication adds an exact copy of a solution from the population of solutions back onto the list. Solutions are chosen for replication with the preferential search algorithm. Replication therefore can neither produce a novel solution nor remove one, but it gently sways the ecology of the population by increasing the popularity of favorable existent solutions. This represents an advocacy of a particular solution under discussion.

Random point mutation. Random point mutation adds a copy of a solution with point mutations, flipping of bits at each aspect of a problem with a probability p_m. The solution on which the operator acts is chosen from the active population with a preferential search algorithm (discussed in more detail below). This represents an attempt of making random changes to the existing ideas, reflected in asking "what if" questions. Random point mutations help escape local maxima of a utility function in the problem space when a utility function is nonlinear and many-peaked.

Intelligent point mutation. A solution is selected from the population with a preferential search algorithm. It makes several (r_m) offspring of the parent solution and selects that of the highest perceived fitness for addition to the population. This represents a proposal of an improved idea derived from existing ideas under discussion. The intelligent point mutation can be useful in maximizing a utility function with one maximum by climbing monotone gradients, but it may perform poorly in a complex utility landscape.

Recombination. Recombination chooses one solution at random and one with a preferential search algorithm. It then creates two offspring from the two parent solutions. Sexual reproduction is simulated with a multiple point cross-over recombination: parent solutions are aligned by aspects, for each of which there is a probability p_s of switching their contents. Of the two offspring, that of higher perceived utility is selected and added to the population. This represents a creation of a new idea from two existing ideas.

Subtractive selection. The preferential search algorithm is used to find the solution with the worst fitness, whereupon it is singled out and deleted from the population. This represents a criticism against a bad idea. Subtractive selection is the only operator that reduces the number of existing solutions and is therefore essential to groups attempting to attain convergence in the population distribution.

Random generation. Finally, random generation of solutions adds a randomly generated solution to the population. There is no use of an individual's utility function, nor any connection to the existing solutions "on the table" at that time. New solutions are generated utterly randomly. This represents a sudden inspiration of a totally unique idea that is unrelated to the existing ideas under discussion.

2.4 Simulation Settings

The following parameter settings were held constant for all simulations: group size N

= 6; problem space dimensionality $M = 10$; number of sample solutions in the preferential search algorithm $r_p = 5$; number of offspring generated in the intelligent point mutation $r_m = 5$; random mutation rate per bit $p_m = 0.2$; probability of random switching in recombination $p_s = 0.4$; number of iterations $t = 60$. It was also assumed that groups were initialized with four random ideas. For each group, the noise parameter v and the bias parameter β were varied from 0 to 1.2 by increments of 0.2.

2.5 Metrics of Group Performance

We use two separate performance metrics: the true utility of the mode solution at the end of group simulation and the convergence of solutions. Convergence is based on entropy

$$H = -\sum_{i=1}^{n} p(x_i) \cdot \log_2 p(x_i) ,\qquad\qquad (3)$$

where $p(x_i)$ is a normalized frequency of the i-th type of solutions in the population. Since the maximum possible value of H is M (this is the case when there are exactly 2^M solutions in the population which are different from each other), $M - H$ is a quantitative measure that intuitively means the number of aspects of the problem on which the group has formed a cohesive opinion. For normality, we will use $(M - H) / M$ as the metric.

3 Results

We first conducted a within-group analysis examining effects of heterogeneity in world views (utility functions) within a group. Here we assumed group members were "balanced" behaviorally; in each iteration, they randomly chose one of the six operators with equal probability. Figure 1 indicates the results with several different settings of within-group variation v and group-level bias β, plotting them in a 2-D performance space using the two metrics described above.

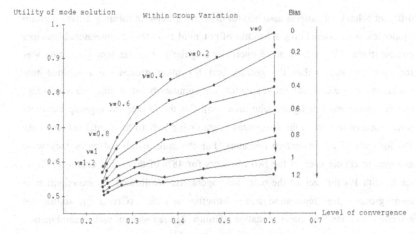

Figure 1: Simulation results showing the effects of within-group heterogeneity and group level bias. The level of convergence and the true utility value of mode solutions for several different noise levels are plotted. $V = 0$ represents the case of completely homogeneous groups, while larger values of V represent more heterogeneous group cases and larger values of bias represent large discrepancies between group utility functions and true utility function.

Group-level bias affects the utility of group solutions while convergence is largely unaffected. On the other hand, within-group variation degrades both convergence and utility. Groups performed better in both performance metrics when they were homogeneous in their utility functions. As the groups' members become more heterogeneous, the true utility value of the mode solution decreased and the final population of solutions after discussion became more diverse. The decrease of the true utility value was particularly drastic; in nonbiased conditions with no heterogeneity, the groups were able to find nearly perfect solutions for the problem (i.e., the utility close to 1). As the groups become more heterogeneous the utility achieved dropped to just above 0.5, meaning that there was no net improvement achieved during the group discussion. This was due to the conflicts of interest among the group members.

Contrary to other findings regarding heterogeneous groups outperforming homogeneous groups on creative and intellectual problem solving tasks [3, 6], our findings indicate the opposite, which may seem to support the negative relationship reported between both surface-level (i.e., demographic) and deep-level (i.e., psychological) diversity and group functioning and performance [2]. We must note here, however, that the diversity considered in this set of experiments is about the individual utility functions only, and not about the individual behavioral patterns.

In order to explore the effects of various compositions of individual behaviors, we ran another set of experiments using the same simulation model with different behavioral patterns assumed for different groups. In forming different group properties, we modeled only a handful of potential evolutionary operators/behaviors combinations. We modeled some operators singularly (e.g., random generation was the only operator within the group), and for other groups we combined two evolutionary operators to reflect increasing complexity of group behavior (e.g., recombination and intelligent point mutation). For the former cases group members were assumed to choose the designated operator for 95% of their total actions, with 1% for each of the other five operators. For the latter combined cases they were assumed to choose each of the two operators for 48% of their total actions (96% in total), with 1% for each of the other four operators. We limited our examination to eight group types: replication and subtractive selection (Group 1); subtractive selection and random point mutation (Group 2); replication and recombination (Group 3); recombination (Group 4); recombination and intelligent point mutation (Group 5); intelligent point mutation and random generation (Group 6); random generation (Group 7); and, finally, the balanced team we used in the previous experiment as a control (Group 0).

Figure 2 shows the results of the second set of experiments comparing group performances with different group properties, plotting them in the same 2-D performance space as used for Figure 1. The effect of group-level bias is similar to that seen in Figure 1. Among the groups examined, the "balanced" Group 0 case was the best in terms of the utility value of the mode solution. Interestingly, however, we saw a variety of different group performances achieved by groups with different properties, seen as a kind of "wave front" near the upper-right corner of the performance space.

We further noticed in Figure 2 that the groups that sit along this wave front were arranged roughly in the order of the balance between selection and variation in evolutionary operators; Group 1, which was the best in terms of the convergence but poor in the mode selection utility, used the combination of replication and subtractive selection, which are both selection-oriented operators. Group 2, the second best in convergence and second worst in mode selection utility within the wave front, used the combination of subtractive selection and random point mutation, which is more variation-oriented than Group 1. Along the way toward Group 0, we saw Group 3 (replication and recombination), Group 4 (recombination only), and Group 5 (recombination and intelligent point mutation), where the qualitative shift of balance of evolutionary operators from selection-oriented to variation-oriented can be seen. It is also notable that the random generation operators (used in Groups 6 and 7) were generally not working for improving group performance.

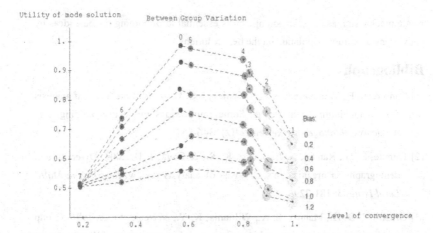

Figure 2: Simulation results showing the effects of group-level difference of behavioral patterns. The level of convergence and the true utility value of mode solutions under varying conditions of group-level behavioral patterns and bias are plotted. Group 0 is the balanced team that uses all of the six evolutionary operators with equal probability; Group 1 uses replication and subtractive selection mostly; Group 2 subtractive selection and random point mutation; Group 3 replication and recombination; Group 4 recombination only; Group 5 recombination and intelligent point mutation; Group 6 intelligent point mutation and random generation; and Group 7 random generation only.

4 Conclusion

The application of the evolutionary paradigm is illuminating for studying collective decision making dynamics because it allows the researcher to remove themselves from the traditional teleology adopted in most simulations of human groups. We have portrayed group dynamics in a novel way by treating members of the group as constituents of an evolutionary environment in which populations of solutions evolve. In this new framework, we have characterized the properties of the population of solutions after discussion as quantitative metrics of the performance of a group. We demonstrated through simulations that heterogeneous groups with random variations in individual utility functions had a drop in both utility and convergence of solution populations compared to more homogeneous groups. We also demonstrated that variations in the compositions of individual behavioral patterns between groups resulted in a large spectrum of performance, in which groups well balanced between reductive and creative evolutionary forces yielded solutions that were highly adaptive by both performance metrics. All operators have a particular utility in appropriate circumstances, but we highlight that recombination operators are particularly important in that they demonstrate creative changes on large and small scales with a single mechanism, as are selection operators essential to promoting the best ideas by converging a solution population on the best solutions.

Bibliography

[1] Dansereau, F., Yammarino, F. J., & Kohles, J. C. 1999. Multiple levels of analysis from a longitudinal perspective: Some implications for theory building. *The Academy of Management Review*, 24(2): 346-357.

[2] Dionne, S. D., Randel, A. E., Jaussi, K. S., & Chun, J. U. 2004. Diversity and demography in organizations: A levels of analysis review. *Research in Multi-Level Issues* 3: 181-229.

[3] Gruenfeld, D. H., Mannix, E. A., Williams, K. Y., & Neale, M. A. 1996. Group composition and decision making: How member familiarity and information distribution affect process and performance. *Organizational Behavior and Human Decision Processes* 67: 1-15.

[4] Kerr, N. L., & Tindale, R. S. 2004. Group performance and decision making. *Annual Review of Psychology*, 55: 623-655.

[5] Klein, M., Faratin, P., Sayama, H., & Bar-Yam, Y. 2006. An annealing protocol for negotiating complex contracts. In J.-P. Rennard, ed., *Handbook of Research on Nature Inspired Computing for Economics and Management*, vol. 2, Chapter XLVIII. Idea Group Publishing.

[6] Nemeth, C. J. 1992. Minority dissent as a stimulant to group performance. In S. Worchel, W. Wood, & J. Simpson, eds., *Group process and productivity*, pp.95-111. Sage.

[7] Yammarino, F. J., Dionne, S. D., Chun, J. U., & Dansereau, F. 2005. Leadership and levels of analysis: A state-of-the-science review. *The Leadership Quarterly*, 16: 879-919.

Chapter 9

Disrupting Terrorist Networks – A Dynamic Fitness Landscape Approach

Philip V. Fellman, School of Business
Southern New Hampshire University
Shirogitsune99@yahoo.com
Jonathan P. Clemens, Intel Corporation
Jonathan.p.clemens@intel.com
Roxana Wright, Keene State College
Rox_wright@yahoo.com
Jonathan Vos Post, Computer Futures, Inc.
Jvospost2@yahoo.com
Matthew Dadmun, Southern New Hampshire University
mangell68@hotmail.com

1.1 Introduction

The study of terrorist networks as well as the study of how to impede their successful functioning has been the topic of considerable attention since the odious event of the 2001 World Trade Center disaster. While serious students of terrorism were indeed engaged in the subject prior to this time, a far more general concern has arisen subsequently. Nonetheless, much of the subject remains shrouded in obscurity, not the least because of difficulties with language and the representation or translation of names, and the inherent complexity and ambiguity of the subject matter.

One of the most fruitful scientific approaches to the study of terrorism has been network analysis (Krebs, 2002; Carley, 2002a; Carley and Dombroski, 2002; Butts, 2003a; Sageman, 2004, etc.) As has been argued elsewhere, this approach may be particularly useful, when properly applied, for disrupting the flow of communications (C^4I) between levels of terrorist organizations (Carley, Krackhardt and Lee, 2001; Carley, 2002b; Fellman and Wright, 2003; Fellman and Strathern, 2004; Carley et al, 2003; 2004). In the present paper we examine a recent paper by Ghemawat and Levinthal, (2000) applying Stuart Kauffman's NK-Boolean fitness landscape approach to the formal mechanics of decision theory. Using their generalized NK-simulation approach, we suggest some ways in which optimal decision-making for terrorist networks might be constrained and following our earlier analysis, suggest ways in which the forced compartmentation of terrorist organizations by counter-terrorist security organizations might be more likely to impact the quality of terrorist organizations' decision-making and command execution.

1.2 General Properties of Terrorist Networks

Without attempting to be either exhaustive or exhausting, recent research on terrorism has revealed several relevant characteristics of terrorist organizations which a prudent modeler ought to keep in mind. These networks are first and foremost, *covert,* which means that they have hidden properties, and our information about them is necessarily incomplete, hence demanding complex methodological tools for determining the properties of the network structure (Butts, 2001, 2003a; Carley 2002a, 2003; Krebs, 2001, Clemens and O'Neill, 2004). While we are primarily concerned in the present paper with formal properties of terrorist networks, it does bear keeping in mind that at the operational level they are *purposive,* which lends them not only formal characteristics, but depending upon the organization in question, a considerable ideological history (Hoffman, 1997; Hoffman and Carr, 1997), and in some cases, rather serious (path-dependent) constraints on recruiting (Codevilla, 2004a; Fellman and Strathern, 2004) targets, and methods (Sageman, 2004). Some other, rather interesting properties of terrorist networks include the fact that they are often separated by larger than normal degrees of distance between their participants, a condition arising from their covert nature (Krebs, 2001; Fellman and Wright, 2003; Carley, 2003). Curiously, this kind of structure appears to have an emergent shape, which can be mapped as a distributed network (Krebs, 2001; Fellman and Strathern, 2004; Clemens and O'Neill, 2004), commonly illustrated by a social network diagram of the 9-11 Hijackers and informally referred to as "the dragon".

Carley et al (2001, 2002b, 2003, 2004) have developed useful models for distinguishing cohesive vs. adhesive organizations as well as defining probably outcomes for the removal of higher visibility nodes. Formal models of network analysis can also suggest where removal of key nodes or vertices can disrupt the organization's ability to transmit commands across hierarchical levels of the organization, thus leading to command degeneration (Butts, 2003a; Carley et al, 2004). The difficulty with this approach is that an important aspect of the dynamics of terrorist networks is that they are learning organizations (Hoffman, 1997; Tsvetovat and Carley, 2003).

If one bears all of these features in mind, some of the complexities of dealing with terrorist organizations become immediately apparent. Terrorists are slippery foes, they are hidden, they have redundant command structures, they change their membership (not all of which changes are visible) and they learn from their mistakes. Nobody who has to deal with terrorist threats wants to see those threats, and the organizations that make them, evolve. The obviousness of this proposition is evidenced by the U.S. reaction to 9-11. What then, are the possible approaches?

Complexity science has afforded a number of approaches to evolution in general (Kauffman, 1993, 1996, 2000) as well as to the evolution of organizations and the ways in which complexity science may be applied to problems of organizational behavior. In particular, Kauffman's NK-Boolean fitness landscape model appears to offer a number of fruitful heuristics (Lissack, 1996; McKelvey, 1999; Meyer, 1996; Fellman et al, 2004). In 1999, seeking to define the formal properties of an optimal business organization decision-making process, Pankaj Ghemawat of Harvard Business School and Daniel Levinthal of the Wharton School ran an agent based simulation of decision-making in order to define the ways in which decisional interdependence and the interdependence of business units affect overall performance (fitness). In the section which follows, we will explore a number of their findings and suggest how they might be applied to inhibiting the fitness of terrorist organizations.

2.1 The Structure of the Ghemawat-Levinthal NK Simulation

A primary goal of the simulation was to model interdependent choices. Levinthal and Ghemawat focus on this aspect of decision making because they are attempting to understand the formal structure of decision-making in organizations with interdependent parts. While they rapidly come to focus on the same measures that we have seen used to characterize terrorist networks, hierarchy and centrality (Butts, 2001, 2003; Carley et al, 2001; 2003, 2004; Carley, 2002, 2003), plus an additional factor of randomness (which most of us are wont to deny) they come at these factors from a slightly different approach than what one might anticipate. N and K are chosen simply as (a) the number of total decisions modeled and (b) the number of decisions which depend upon other decisions. As they explain (p. 16):

> The model has two basic parameters, N, the total number of policy choices and K ($<$ N), the number of policy choices that each choice depends upon. More specifically, each of the choices is assumed to be binary, and choice-by-choice contributions to fitness levels are drawn randomly from a uniform distribution over [0,1] for each of the $2K+1$ distinct payoff-relevant combinations a choice can be part of. Total fitness is just the average of these N choice-by-choice fitness levels. Note that with K equal to its minimum value of 0, the fitness landscape is smooth and single-peaked: changes in the setting of one choice variable do not affect the fitness contributions of the remaining N-1 choice variables. At the other extreme, with K equal to N-1, a change in a single attribute of the organism or organization changes the fitness contribution of all its attributes, resulting in many local peaks rather than just one, with each peak associated with a set of policy choices that have some internal consistency. No local peak can be improved on by

perturbing a single policy choice, but local peaks may vary considerably in their fitness levels.

However, a pure NK approach suffers from the disadvantage that all choices are assumed to be equal. To avoid this problem and to model a richer decisional landscape they employ an adjacency matrix, moving us into the familiar Carter Butts (1997, 2000, 2001, 2003a, 2003b) territory of connected graphs (Carley and Butts, 1997; Butts, 2000), and formal, axiomatically determined complex systems.[1]

2.2 Adjacency Matrices

With respect to this process, in Kauffman's original NK-Boolean dynamic fitness landscape model all of the potential choices that a firm could make were considered to be equally important and the search for higher levels of fitness was carried out through a random walk across the fitness landscape (Meyer, 1996). However, under these conditions the NK model could not account for the asymmetric relationship between strategic choices that decision-makers faced. In order to better represent the asymmetric nature of the choices facing a corporation (or, as in our case, a terrorist organization), Ghemawat and Levinthal replace the interactive parameter K, as described above, with an adjacency matrix:

> How different choices (the vertices in the graph) are linked (the lines in the graph). In such a matrix, choice variable j's effect on other variables is represented by the patters of 0s and 1s in column j, with a value of 1 indicating that the payoff to the variable in the row being considered is contingent on variable j, and a value of 0 denoting independence. Similarly, reading across row i in such a matrix indicates the variables the payoff of choice variable i is itself contingent upon. The principle diagonal of an adjacency matrix always consists of 1's, but the matrix itself need not be symmetric around that diagonal.

In order to simplify their examination of the relationship between asymmetric choices Ghemawat and Levinthal elected to look at two adjacency matrices that

---

[1] Replacement of the interactivity parameter, K, with an adjacency matrix is meant to let us generalize the NK approach in the directions presently of interest. A few general observations can be made about special types of graphs and the fitness landscapes that they induce over the choices and linkages they embody. Thus, given disconnected graphs, fitness landscapes are smooth as the choices corresponding to disconnected vertices are varied—irrespective of the values of other variables. Such vertices therefore lend themselves to the notion of universal (and uncontingent) best practices. And for star graphs, in which one central choice influences the payoffs from each of N-1 peripheral choices but other linkages among choices are absent (corresponding to an adjacency matrix with 1's in the first column and along the principal diagonal and 0s everywhere else), getting the first choice right is sufficient, in conjunction with a standard process of local search in an invariant environment, to lead the organization to the global optimum. But what about graphs more generally? Exhaustive enumeration of all the graphs with N vertices and analysis of their fitness landscapes is unlikely to prove productive for even moderately large N: the number of 6-vertice graphs is 157, 7-vertice graphs 1,044, 8-vertice graphs 12,346, and so on. Restricting attention to connected graphs doesn't help much with the numbers problem since the number of connected graphs grows much more quickly than the number of disconnected graphs: with N equal to 5, disconnected graphs account for about 38% of the total, but with N equal to 8, that figure is down to less than 10%. We therefore pass up the opportunity to engage in exhaustive (and exhausting) enumeration. We begin, instead, by considering two classes of adjacency matrices that highlight two fundamental sources of asymmetry among choices, in terms of hierarchy and centrality, and comparing them with the canonical NK structure on which previous work has focused. (pp. 17-18).

highlight the classical types of choice asymmetry: hierarchy, and centrality. In the hierarchical matrix choice 1 is the most important influencing all other choices below it choice two is the second most important and so on to the final choice (in this case choice 10 were N=10) which is influences by all proceeding choices but influences only itself. For the centrality matrix choice 1 is the most central both influencing and also being influenced by all other possible choices, choice two is the second most central being influenced by all other choices and influencing all choices with exception 10 and so on. These two matrices are benchmarked against a traditional NK structure. The matrix in this case is structured such that there will be K 1's in each row and column but they will be randomly distributed across the matrix.[2] In their simulation, Ghemawat and Levinthal put K=6 which proved the same number of peaks at the other two matrices [Ghemawat and Levinthal 2005].

2. 3 Hierarchy

Ghemawat and Levinthal treat hierarchical decisions as directed trees where the 1 appears to the left of the principal diagonal. In this regard as we have explained above, choice 1 is the most hierarchically important, choice 2 the second most important, etc. (Ghemawat and Levinthal, 2000).

Figure 4a. Hierarchy

	1	2	3	4	5	6	7	8	9	10
1	1	0	0	0	0	0	0	0	0	0
2	1	1	0	0	0	0	0	0	0	0
3	1	1	1	0	0	0	0	0	0	0
4	1	1	1	1	0	0	0	0	0	0
5	1	1	1	1	1	0	0	0	0	0
6	1	1	1	1	1	1	0	0	0	0
7	1	1	1	1	1	1	1	0	0	0
8	1	1	1	1	1	1	1	1	0	0
9	1	1	1	1	1	1	1	1	1	0
10	1	1	1	1	1	1	1	1	1	1

[2] In most cases Ghemawat and Levinthal use a Poisson distribution or another, uniform distribution, noting, in any case that the probability distribution is not likely to be a mathematically relevant factor in the overall distribution of decision outcomes (i.e., the probability distribution is not the determinative property).

2. 4 Centrality

In contrast, their treatment of centrality involves interconnected decisions and, hence produces an almost perfect 90 degree rotated distribution (Ghemawat and Levinthal, 2000):[3]

Figure 4b. Centrality

	1	2	3	4	5	6	7	8	9	10
1	1	1	1	1	1	1	1	1	1	1
2	1	1	1	1	1	1	1	1	1	0
3	1	1	1	1	1	1	1	1	0	0
4	1	1	1	1	1	1	1	0	0	0
5	1	1	1	1	1	1	0	0	0	0
6	1	1	1	1	1	1	0	0	0	0
7	1	1	1	1	0	0	1	0	0	0
8	1	1	1	0	0	0	0	1	0	0
9	1	1	0	0	0	0	0	0	1	0
10	1	0	0	0	0	0	0	0	0	1

3. 1 Modeling Policy Choices

Levinthal and Ghemawat then benchmark what happens in these two types of structures against the random (but symmetric) activity which is built into the canonical NK structure. As they explain:

> For all three structures, an organization's policy choices are represented by a vector of length N where each element of the vector can take on a value of 0 or 1 (not to be confused with the 0s and 1s assigned, respectively, to denoting the absence or presence of linkages between every pair of policy elements). The overall fitness landscape will then consist of 2^N possible policy choices, with the overall behavior of the organization characterized by a vector $\{x_1, x_2, \ldots, x_N\}$, where each x_i takes on the value of 0 or 1.

---

[3] (p. 19) "The particular form of hierarchy we explore in this paper has 1s as all the entries to the left of the principal diagonal (see Figure 4a). Choice 1 is hierarchically the most important, choice 2 the second most important, and so on. In contrast, in a set of interaction patterns ordered by a centrality measure, policies vary in terms of their interdependence with other policy choices and this interdependence is taken to be symmetric (to distinguish it as sharply as possible from the one-way influences of hierarchy). As a result, the 1s to the left of the principal diagonal are mirrored by 1s to its right. Whether the 1s cluster centrally in the adjacency matrix, however, depends on the order in which choice variables are labeled. The particular form of centrality we explore in this paper embodies a structure and a labeling scheme that has 1s as all the entries to the left of the inferior diagonal (but distributed symmetrically to the left and the right of the principal diagonal)—see Figure 4b. Thus, choice 1 is most central, choice 2 second most central, and so on.

If the contribution of a given element, xi, of the policy vector to the overall payoff is influenced by Ki other elements—in ways that vary across the three structures we will analyze—then it can be represented as $f(xi|xi1, xi2, \ldots, xiKi)$. Therefore, each element's payoff contribution can take on $2Ki+1$ different values, depending on the value of the attribute itself (either 0 or 1) the value of the Ki other elements by which it is influenced (each of these Ki values also taking on a value of 0 or 1) and—less commonly highlighted —the luck of the draw. Specifically, it is common to assign a random number drawn from the uniform distribution from zero to one to each possible $f(xi|xi1, xi2, \ldots, xKi)$ combination with the overall fitness value then being defined as
$\Sigma i=1$ to N $f(xi|xi1, xi2, \ldots, xiKi)$ / N. (pp. 19-20)

Their simulation structure assumes for the random benchmark that K=6, primarily because this value generates roughly the same number of local peaks as the hierarchical and central distributions.[4] Similarly they set N = 10, which is sufficient to generate more than a million distinct graphs, which allows them to report results averaged over a thousand independent landscapes which share the same structure. These landscapes will be either hierarchical, central or random, characterized by the particular adjacency matrix structure for each type, but with a distinct seeding (0,1) from a uniform random distribution for the fitness of the policy variables.

4.1 Analytical Results of the Ghemawat-Levinthal Simulation

The first question which the authors ask "what are the effects of presetting a certain number of policy choices equal to their values at the global optimum with the remaining choices determined by a process of local search?" is interesting from a complexity science point of view, but not immediately obvious in its application to terrorist organizations. The reason for this is that while answering this question allows Ghemawat and Levinthal to address issues of strategic planning and "grand strategy" in business organizations, it doesn't really provide a reliable guide for the C⁴I functioning of terrorist organizations. If this were all that their simulation achieved, it would have rather limited interest for us. However their second question "what happens when one of the N values of the policy variable is preset to a value inconsistent with the value of that variable for the global peak?" is of very substantial interest as it speaks to exactly the kind of distortions which we would wish to induce in the terrorist decision-making chain.

---

[4] Ghemawat and Levinthal provide five additional caveats, of which the three important for our purposes are: "A number of important assumptions, based on prior applications, are built into this specification. First of all, there is the emphasis on choice under uncertainty. In addition to its arguable descriptive realism, initial uncertainty helps explain why an organization launched over a fitness landscape may not instantly alight on the globally optimal policy vector. Second, there is the assumption that randomness takes the form of a uniform distribution. While some might argue that this distribution is too diffuse, we retain this assumption to provide at least some basis for numerical comparability with prior work, which suggests, among other things, that the structure of the fitness landscape is not sensitive to the particular probability distribution employed (Weinberger, 1991). Third, there is the equal weighting of different choices in terms of their direct contribution (potential) to overall fitness. Again, we retain this prior assumption even though we intend to focus on asymmetries among choices. Putting different weights on the direct contributions of choice elements does not seem to us to be the best way of gaining insight into the indirect contributions that choice elements can make to overall performance by virtue of the linkages among them. (p. 21)

The first interesting result of the simulation is the difference between an optimal preset of policy configurations, the hierarchical, the central and the random simulation arrays:[5]

Figure 5. Value of Partially Articulated Activity Maps

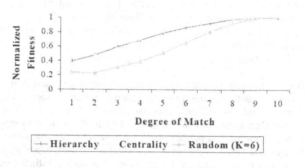

Figure 6. Partially Specified Activity Maps and Proximity to Global Optimum

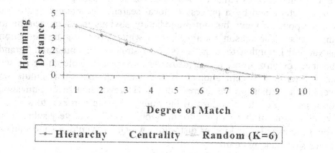

Ghemawat and Levinthal describe this as:

> With a degree of match of 1, only the first, most strategic, variable is set equal to the global optimum. As more variables are matched with their settings at the global optimum, fitness rises steadily according to **Figure 5.** However, the global optimum is not approached until nearly all policy variables are specified to equal their settings at the global optimum. Similarly, hamming distances tend to be quite large (see **Figure 6**).
> The gap between the curve depicting performance under the random network structure and the other two curves indicates the power of presetting more strategic variables to their values at the global optimum. In contrast, the gap between the

---

[5] Ibid.

realized fitness level and the value of 1, indicates the loss from the not fully articulating the optimal policy array.

Admittedly, while this is an interesting result, so far it does not tell us much more than the intuitively plausible idea that an inability or failure to articulate global optima means that you probably never get there. There is, however, a slightly more interesting subtext here with respect to the complexity of rugged fitness landscapes, something originally articulated by Farmer, Packard and Kauffman ("The Structure of Rugged Fitness Landscapes" in Kauffman, 1993):

> To make more sense of these patterns, it is useful to note that the fitness landscapes we are analyzing are quite complex, typically comprising over 40 local peaks. In such worlds, the powers of local search are relatively limited. Local search rapidly leads to the identification of a local peak but conveys no assurance about the local peak's global properties (i.e., its fitness value relative to the global optimum). Presetting the most strategic variables to their values at the global optimum does lead to the identification of a better-than-average local peak (recall that the normalized fitness value would have a value of zero if the average realized fitness level equaled the average value of local peaks in the fitness landscape). However, a high level of specificity is necessary to obtain the highest possible fitness levels or configurations close to the global optimum: in rugged landscapes, there are just too many positive-gradient paths that lead to local peaks other than the global one. (p. 26)

What is perhaps most interesting here are the suggestions that (1) the events of 9-11 are likely to prove the exception, rather than the rule (also giving some optimism to the long-run possibility of ultimately negating terrorist threats, something as unthinkable today as the end of the Cold War was forty or fifty years ago) and (2) the indication that even very hierarchical terrorist organizations are unlikely to reach an optimal policy set through local search (and, following Kauffman, there simply are no other search mechanics available). Figure 7 provides what may be a slightly more interesting insight. Much in the fashion of Lissack (1996), Ghemawat and Levinthal use an approach similar to "patching" in order to simulate a feedback situation where initial errors in policy choice can be corrected and the corrections incorporated into subsequent searches. They note that (p. 26):

> ...while the articulation of and insistence on adherence to a single (or low-dimensional) strategic choice may not be sufficient to lead to the identification of a high-performing set of choices, a lack of such strategic discipline is likely to lead to even less attractive results. Compare the top line in **Figure 7**, tracing the value of partially articulated activity maps in a hierarchical context in which preset choices cannot be varied (á la **Figure 5**) with the bottom line, which looks at a hierarchical context in which the preset policy choices *can* be revised in the process of local search. It turns out that with the degree of match of 1, the latter, "unconstrained" approach underperforms the "constrained" approach, and the gap between the two widens for intermediate degrees of match prior to convergence as the degree of match hits 10. Similarly, the unconstrained approach fails to generate smaller hamming distances than the constrained approach. In that sense, strategic discipline *is* useful.

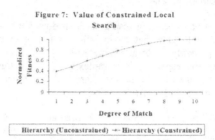

Figure 7: Value of Constrained Local Search

In this regard, as we have previously suggested (Fellman, Wright and Sawyer, 2003; Fellman and Wright, 2003) forcing compartmentation of terrorist cells and interrupting communications between command echelons is likely to significantly impede the ability of those cells to function effectively. From a counter-terrorism perspective, however, the most interesting part of the simulation is the section which deals with "the constraints of history". In this case, Ghemawat and Levinthal artificially inject a variable inconsistent with the global maximum and then observe how the three different types of organization adapt to this distortion. Their Figure 8 summarizes the (normalized) fitness levels achieved in the simulation when they preset a variable to a value which is inconsistent with the global optimum; this is shown below.

Figure 8. Constraints of History

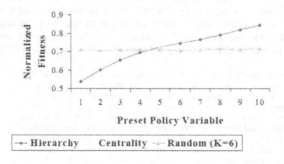

The green line, representing a random pattern of interactions, not surprisingly has a fitness which is essentially "history independent". The more surprising finding is that while under conditions of hierarchy the fitness level changes quite rapidly as the mismatched variable is shifted from the most important decision to decisions of less importance, centrality conveys virtually no "recovery advantage" at all. This is a striking contrast to Porter (1996) who argues that under conditions of "third order strategic fit" (strong interlinkage/strong centrality) the decision chain becomes "as strong as its strongest link". In dealing with terrorist organizations, which are primarily hierarchical in nature (Sageman, 2004), what this finding says is that disinformation is a useful tactic (or strategy) only if it succeeds in influencing one of the key decisional variables. In other words, disinformation at the local level is unlikely to have any lasting impact on terrorist organizations. This finding also challenges the institutional wisdom of assigning case officers in the field to this type of counter-terrorism operation (Gerecht, 2001; Codevilla, 2004b).[6] In fact, from an operational point of view, the hierarchical nature of terrorist organizations means that there may be something of a mismatch in the entire targeting process. As Ghemawat and Levinthal note, "Less central variables not only do not constrain, or substantially influence the payoff of many other choices, but they themselves are not greatly contingent upon other policy choices. Being contingent on other policy choices facilitates compensatory shifts in policy variables other than the one that is preset. As a result of the absence of such contingencies, the presetting of lower-order policy choices is more damaging to fitness levels under the centrality structure." (p. 27) The problem, however, is that decisionmaking in terrorist organizations is apt to be operating under conditions of hierarchy rather then centrality. Perhaps even more annoyingly, those organizations operating against terrorist cells may themselves be organized in a more modern fashion, availing themselves of divisional interlinkages and flat management structure, so that the mismatch between operational objectives and terrorist organizations may, in fact, prove organizationally damaging to the counter-terrorist organization both in a relative and an absolute sense. In this regard, pointless, or fruitless counter-terrorism operations, particularly when conducted at the field level may do considerably more damage than good, particularly to the organization striving to combat terrorism. This is a feature of counter-terrorism that is not entirely unfamiliar to case officers who have operated in this capacity (Gilligan, 2003; Gerecht 2001).

In their final section, Ghemawat and Levinthal examine choice structures.[7] To do this, they set up three classes of variables: independent variables (1-3), whose payoff is not dependent upon that of any other variables, but which influences the payoff of the dependent variables (4-6), and three variables (7-9) which are simply independent of all others (Figure 12).

---

[6] Ghemawat and Levinthal test this another way and come to essentially the same conclusion: "Another striking feature of this set of simulations concerns how few of the optima with preset mismatches constitute local peaks of the fitness landscape. Given the importance of configurational effects, one might reasonably conjecture that constraining one variable to differ from the global optimum would lead to the selection of a different, non-global, peak in the fitness landscape. However, Figure 9 indicates that this is relatively uncommon except as one turns to presetting the least important variables under the hierarchy and centrality structures." (pp. 28-29)

[7] "The broader suggestion is that the 'natural' adjacency matrices we have looked at so far mix up at least three very different types of effects: influence, contingency and autonomy. Variables may be more or less influential to the extent that they affect the payoffs to other variables. In an adjacency matrix, this is represented by the prevalence of 1s in the relevant column. Independent of influence, the payoffs from specific variables may be more or less contingent on other choices, as reflected in the number of 1s in the relevant row of the adjacency matrix. And autonomy is characterized by variables that are neither influential nor contingent: variables that correspond, in graph-theoretic terms, to disconnected vertices. In this subsection, we look at a choice structure—distinct from the three that we have already examined—that distinguishes particularly clearly among these three effects." (p. 29)

Figure 12. Influence, Contingency, and Autonomy

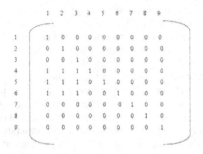

They illustrate their findings in Figure 13:

Figure 13. Constraints of History:
Normalized Fitness Levels

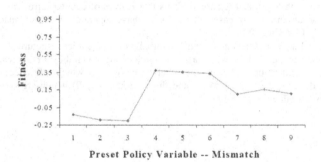

Preset Policy Variable -- Mismatch

Interestingly, Ghemawat and Levinthal are surprised that constraining the independent variables to values different from the global maximum affects overall fitness more than the constraint of contingent variables:

> **Figure 13** indicates that constraining one of the "influential" variables to differ from the global maximum has a profound effect on the relative fitness level of the constrained optimum. Surprisingly, constraining the independent variables to differ from the global optimum has a larger impact than constraining the seemingly more important "contingent" variables. The reason for this is that the presence of contingency allows for the possibility of substituting or compensating changes in policy variables. While tightly linked interaction patterns have generally been viewed as fragile, the equifinality that high levels of interaction engender also allows for a certain robustness. In contrast, when an autonomous variable is misspecified, that doesn't create negative ramifications elsewhere in the system of policy choices; at the same time, however, there is no opportunity to compensate for the misspecification.

5. 1 Conclusion

Terrorist networks are complex, possessing multiple, irreducible levels of complexity and ambiguity. This complexity is compounded by the covert, dynamic nature of terrorist networks where key elements may remain hidden for extended periods of time and the network itself is dynamic. The NK-Boolean fitness landscape simulation approach offers a number of tools which may be particularly useful in sorting out some of the complexities of terrorist networks. In the present study the biggest surprise was that coupled decisional structures have a compensatory feature which makes them less vulnerable to certain kinds of decisional disruption than hierarchical decision structures. The lesson here for intelligence organizations is that in dealing with terrorist networks, counter-terrorism efforts are best be directed at hitting them hard, hitting them at the highest levels of the hierarchy and leaving the "cleanup" to local authorities or local law enforcement organizations. In this context, there are many circumstances where less engagement may, in fact, yield more results. Conversely, if "sometimes less is more", clearly "sometimes more is a whole lot less." In other words, it's better to do a little of the right thing, or even do nothing at all than to undertake counter-productive activities on a massive scale.

References

Butts, Carter T. (2000). "An Axiomatic Approach to Network Complexity." Journal of Mathematical Sociology, 24(4), 273-301.

Butts, Carter T. (2001). "The Complexity of Social Networks: Theoretical and Empirical Findings." Social Networks, 23(1), 31-71

Butts, Carter T. (2003a) "Network Inference, Error, and Informant (In)Accuracy: A Bayesian Approach", Social Networks 25(2) 103-30.

Butts, Carter T. (2003b). "Predictability of Large-scale Spatially Embedded Networks." In Ronald Breiger, Kathleen Carley, and Philippa Pattison (eds.), Dynamic Social Network Modeling and Analysis: Workshop Summary and Papers, 313-323. Washington, D.C.: National Academies Press.

Carley, Kathleen M. and Butts, Carter T. (1997). "An Algorithmic Approach to the Comparison of Partially Labeled Graphs." In Proceedings of the 1997 International Symposium on Command and Control Research and Technology. June. Washington, D.C.

Carley, Kathleen; Lee, Ju-Sung; and Krackhardt, David; (2001) "Destabilizing Networks", Connections 24(3): 31-34, INSNA (2001)

Carley, Kathleen (2002a) Modeling "Covert Networks", Paper prepared for the National Academy of Science Workshop on Terrorism, December 11, 2002

Carley, Kathleen (2002b) "Inhibiting Adaptation", Proceedings of the 2002 Command and Control Research and Technology Symposium, Naval Postgraduate School, Monterey, CA (2002)

Carley, Kathleen; Dombroski, Matthew; Tsvetovat, Maksim; Reminga, Jeffrey; Kamneva, Natasha (2003) "Destabilizing Dynamic Covert Networks", Proceedings of the 8th International Command and Control Research and Technology Symposium, National Defense War College, Washington, D.C. (2003)
http://www.casos.cs.cmu.edu/publications/resources_others/a2c2_carley_2003_d estabilizing.pdf

Carley, Kathleen; Diesner, Jana; Reminga, Jeffrey; and Tsvetovat,, Maksim (2004), "Toward an end-to-end approach for extracting, analyzing and visualizing network data", ISRI, Carnegie Mellon University (2004)

Clemens, Jonathan P. and O' Neill, Lauren "Discovering an Optimum Covert Network", Santa Fe Institute, Summer, 2004

Codevilla, Angelo (2004a) "Doing it the Hard Way", Claremont Review of Books, Fall, 2004 http://www.claremont.org/writings/crb/fall2004/codevilla.html

Codevilla, Angelo (2004b) "Why U.S. Intelligence is Inadequate and How to Fix It", Center for Security Policy, Occasional Papers Series, December, 2004
http://www.centerforsecuritypolicy.org/occasionalpapers/Why-US-Intelligence-Is-Inadequate.pdf

Fellman, Philip V.; Sawyer, David; and Wright, Roxana (2003) "Modeling Terrorist Networks - Complex Systems and First Principles of Counter-Intelligence," Proceedings of the NATO Conference on Central Asia: Enlargement, Civil – Military Relations, and Security, Kazach American University/North Atlantic Treaty Organization (NATO) May 14-16, 2003

Fellman, Philip V. and Wright, Roxana (2003) "Modeling Terrorist Networks: Complex Systems at the Mid-Range", paper prepared for the Joint Complexity Conference, London School of Economics, September 16-18, 2003
http://www.psych.lse.ac.uk/complexity/Conference/FellmanWright.pdf

Fellman, Philip V. and Strathern, Mark (2004) "The Symmetries and Redundancies of Terror: Patterns in the Dark", Proceedings of the annual meeting of the North American Association for Computation in the Social and Organizational Sciences, Carnegie Mellon University, June 27-29, 2004.
http://casos.isri.cmu.edu/events/conferences/2004/2004_proceedings/V.Fellman, Phill.doc

Gerecht, R,.M., "The Counterterrorist Myth", The Atlantic Monthly, July-August, 2001

Ghemawat, Pankaj and Levinthal, Daniel (2000) "Choice Structures, Business Strategy and Performance: A Generalized NK-Simulation Approach", Reginald

H. Jones Center, The Wharton School, University of Pennsylvania (2000) http://www.people.hbs.edu/pghemawat/pubs/ChoiceStructures.pdf

Gilligan, Tom CIA Life: 10,000 Days with the Agency, Intelligence E-Publishing Company (November 2003)

Hoffman, Bruce (1997) "The Modern Terrorist Mindset: Tactics, Targets and Technologies", Centre for the Study of Terrorism and Political Violence, St. Andrews University, Scotland, October, 1997

Hoffman, Bruce and Carr, Caleb (1997) "Terrorism: Who Is Fighting Whom?" World Policy Journal, Vol. 14, No. 1, Spring 1997

Kauffman, Stuart (1993), The Origins of Order, Oxford University Press (Oxford and New York: 1993)

Kauffman, Stuart (1996), At Home In the Universe, Oxford University Press (Oxford and New York: 1996)

Kauffman, Stuart (2000) Investigations, Oxford University Press (Oxford and New York, 2000)

Krebs, Valdis (2001) "Uncloaking Terrorist Networks", First Monday, (2001) http://www.orgnet.com/hijackers.html

Lissack, Michael, (1996) "Chaos and Complexity: What Does That Have to Do with Knowledge Management?", in Knowledge Management: Organization, Competence and Methodology, ed. J. F. Schreinemakers, Ergon Verlog 1: 62-81 (Wurzburg: 1996)

McKelvey, Bill (1999) "Avoiding Complexity Catastrophe in Coevolutionary Pockets: Strategies for Rugged Landscapes", Organization Science, Vol. 10, No. 3, May–June 1999 pp. 294–321

Meyer, Chris "What's Under the Hood: A Layman's Guide to the Real Science", Center for Business Innovation, Cap Gemini, Ernst and Young, Conference on Embracing Complexity, San Francisco, July 17-19, 1996

Porter, Michael (1996) "What is Strategy?", Harvard Business Review, November-December, 1996

Sageman, Mark (2004) Understanding Terror Networks, University of Pennsylvania Press, 2004.

Tsvetovat, Max & Carley, Kathleen. (2003). Bouncing Back: Recovery mechanisms of covert networks. NAACSOS Conference 2003

Chapter 10

The World as Evolving Information

Carlos Gershenson[1,2,3]

[1] Computer Sciences Department, Instituto de Investigaciones en
Matemáticas Aplicadas y en Sistemas
Universidad Nacional Autónoma de México[1]
Ciudad Universitaria, A.P. 20-726, 01000 México D.F. México
cgg@unam.mx http://turing.iimas.unam.mx/ cgg
[2] New England Complex Systems Institute
238 Main Street Suite 319 Cambridge, MA 02142, USA
[3]Centrum Leo Apostel, Vrije Universiteit Brussel
Krijgskundestraat 33 B-1160 Brussel, Belgium

This paper discusses the benefits of describing the world as information, especially in the study of the evolution of life and cognition. Traditional studies encounter problems because it is difficult to describe life and cognition in terms of matter and energy, since their laws are valid only at the physical scale. However, if matter and energy, as well as life and cognition, are described in terms of information, evolution can be described consistently as information becoming more complex.

The paper presents eight tentative laws of information, valid at multiple scales, which are generalizations of Darwinian, cybernetic, thermodynamic, psychological, philosophical, and complexity principles. These are further used to discuss the notions of life, cognition and their evolution.

---

[1]Current affiliation. A considerable part of this work was developed while at other institutions.

1 Introduction

Throughout history we have used concepts from our current technology as metaphors to describe our world. Examples of this are the description of the body as a factory during the Industrial Age, and the description of the brain as a computer during the Information Age. These metaphors are useful because they extend the knowledge acquired by the scientific and technological developments to other areas, illuminating them from a novel perspective. For example, it is common to extend the particle metaphor used in physics to other domains, such as crowd dynamics [27]. Even when people are not particles and have very complicated behaviour, for the purposes of crowd dynamics they can be effectively described as particles, with the benefit that there is an established mathematical framework suitable for this description. Another example can be seen with cybernetics [4, 28], where the system metaphor is used: everything is seen as a system with inputs, outputs, and a control that regulates the internal variables of the system under the influence of perturbations from its environment. Yet another example can be seen with the computational metaphor [60], where the universe can be modelled with simple discrete computational machines, such as cellular automata or Turing machines.

Having in mind that we are using metaphors, this paper proposes to extend the concept of information to describe the world: from elementary particles to galaxies, with everything in between, particularly life and cognition. There is no suggestion on the nature of reality as information [58]. This work only explores the advantages of *describing* the world as information. In other words, there are no ontological claims, only epistemological.

In the next section, the motivation of the paper is presented, followed by a section describing the notion of information to be used throughout the paper. In Section 4, eight tentative laws of information are put forward. These are applied to the notions of life (Section 5) and cognition (Section 6). The paper closes presenting future work and conclusions.

2 Why Information?

There is a great interest in the relationship between energy, matter, and information [32, 54, 43]. One of the main reasons for this arises because this relationship plays a central role in the definition of life: Hopfield [30] suggests that the difference between biological and physical systems is given by the meaningful information content of the former ones. Not that information is not present in physical systems, but—as Roederer puts it—information is *passive* in physics and *active* in biology [49]. However, it becomes complicated to describe how this information came to be in terms of the physical laws of matter and energy. In this paper the inverse approach is proposed: let us describe matter and energy in terms of information. If atoms, molecules and cells are described as information, there is no need of a *qualitative* shift (from non-living to living matter) while describing the origin and evolution of life: this is translated into a *quantitative* shift (from less complex to more complex information).

There is a similar problem when we study the origin and evolution of cognition [20]: it is not easy to describe cognitive systems in terms of matter and energy. The drawback with the physics-based approach to the studies of life and cognition is that it requires a new category, that in the best situations can be referred to as "emergent". Emergence is a useful concept, but it this case it is not explanatory. Moreover, it stealthily introduces a dualist view of the world: if we cannot relate properly matter and energy with life and cognition, we are forced to see these as separate categories. Once this breach is made, there is no clear way of studying or understanding how systems with life and cognition evolved from those without it. If we see matter and energy as particular, simple cases of information, the dualist trap is avoided by following a continuum in the evolution of the universe. Physical laws are suitable for describing phenomena at the physical scale. The tentative laws of information presented below aim at being suitable for describing phenomena *at any scale*. Certainly, there are other approaches to describe phenomena at multiple scales, such as general systems theory and dynamical systems theory. These approaches are not exclusive, since one can use several of them, including information, to describe different aspects of the same phenomenon.

Another benefit of using information as a basic descriptor for our world is that the concept is well studied and formal methods have already been developed [14, 46], as well as its philosophical implications have been discussed [19]. Thus, there is no need to develop a new formalism, since information theory is well established. I borrow this formalism and interpret it in a new way.

Finally, information can be used to describe other formalisms: not only particles and waves, but also systems, networks, agents, automata, and computers can be seen as information. In other words, it can contain other descriptions of the world, potentially exploiting their own formalisms. Information is an *inclusive* formalism.

3 What Is Information?

Extending the notion of Umwelt [57], the following notion of information can be given:

Notion 1 *Information is anything that an agent can sense, perceive, or observe.*

This notion is in accordance with Shannon's [52], where information is seen as a just-so arrangement, a defined structure, as opposed to randomness [12, 13], and it can be measured in bits. This notion can be applied to everything that surrounds us, including matter and energy, since we can perceive it—because it has a defined structure—and we are agents, according to the following notion:

Notion 2 *An agent is a description of an entity that acts on its environment [22, p. 39].*

Noticing that agents (and their environments) are also information (as they can be perceived by other agents, especially us, who are the ones who *describe* them as agents), an agent can be a human, a cell, a molecule, a computer program, a society, an electron, a city, a market, an institution, an atom, or a star. Each of these can be described (by us) as *acting* in their environment, simply because they *interact* with it. However, not all information is an agent, e.g. temperature, color, velocity, hunger, profit.

Notion 3 *The environment of an agent consists of all the information* interacting *with it.*

Information will be relative to the agent perceiving it[2]. Information can exist in theory "out there", independently of an agent, but for practical purposes, it can be only spoken about once an agent—not necessarily a human—perceives / interacts with it. The *meaning* of the information will be given by the *use* the agent perceiving it makes of it [59], i.e. how the agent responds to it [7]. Thus, Notion 1 is a *pragmatic* one. Note that perceived information is different from the meaning that an agent gives to it. Meaning is an *active* product of the *interaction* between information and the agent perceiving it [13, 44].

Like this, an electron can be seen as an agent, which perceives other electrons as information. The same description can be used for molecules, cells, and animals. We can distinguish:

First order information is that which is perceived directly by an agent. For example, the information received by a molecule about another molecule

Second order information is that which is perceived by an agent about information perceived by another agent. For example, the information perceived by a human observer about a molecule receiving information about another molecule.

Most of the scientific descriptions about the world are second order information, as we perceive how agents perceive and produce information. The present approach also introduces naturally the role of the observer in science, since everything is "observing" the (limited, first order) information it interacts with from its own perspective. Humans would be second-level observers, observing the information observed by information. Everything we can speak about is observed, and all agents are observers.

Information is not necessarily conserved, i.e. it can be created, destroyed, or transformed. These can take place only through interaction. *Computation* can be seen as the *change* in information, be it creation, destruction, or transformation. Matter and energy can be seen as particular types of information that cannot be created or destroyed, only transformed, along with the well-known properties that characterize them.

---

[2]Shannon's information [52] deals only with the technical aspect of the transmission of information and not with its *meaning*, i.e. it neglects the semantic aspect of communication.

The amount of information required to describe a process, system, object, or agent determines its *complexity* [46]. According to our current knowledge, during the evolution of our universe there has been a shift from simple information towards more complex information [2] (the information of an atom is less complex than that of a molecule, than that of a cell, than that of a multicellular organism, etc.). This "arrow of complexity"[11] in evolution can guide us to explore general laws of information.

4 Tentative Laws of Information

Seeing the world as information allows us to describe general laws that can be applied to everything we can perceive. Extending Darwin's theory [15], the present framework can be used to reframe "universal Darwinism" [17], which explores the idea of evolution beyond biological systems. In this work, the laws that describe the general behaviour of information as it evolves are introduced. These laws are only *tentative*, in the sense that they are only presented with arguments in favour of them, but they still need to be thoroughly tested.

4.1 Law of Information Transformation

Since information is relative to the agents perceiving it, *information will potentially be* transformed *as different agents perceive it*. Another way of stating this law is the following: *information will potentially be transformed by* interacting *with other information*. This law is a generalization of the Darwinian principle of random variation, and ensures *novelty* of information in the world. Even when there might be static information, different agents can perceive it differently and interact with it, potentially transforming it. Through evolution, the transformation of information generates a *variety* or *diversity* that can be used by agents for novel purposes.

Since information is not a conserved quantity, it can increase (created), decrease (destroyed), or be maintained as it is transformed.

As an example, RNA polymerase (RNAP) can make errors while copying DNA onto RNA strands. This slight random variation can lead to changes in the proteins for which the RNA strands serve as templates. Some of these changes will lead to novel proteins that might improve or worsen the function of the original proteins.

The transformation of information can be classified as follows:

Dynamic. Information changes itself. This could be considered as "objective, internal" change.

Static. The agent perceiving the information changes, but the information itself does not change. There is a dynamic change but in the agent. This could be considered as "subjective, internal" change.

Active. An agent changes information in its environment. This could be considered as an "objective, external" change.

Stigmergic. An agent makes an active change of information, which changes the perception of that information by another agent. This could be considered as "subjective, external" or "intersubjective" change.

4.2 Law of Information Propagation

Information propagates as fast as possible. Certainly, only some information manages to propagate. In other words, we can assume that different information has a different "ability" to propagate, also depending on its environment. The "fitter" information, i.e. that which manages to persist and propagate faster and more effectively, will prevail over other information. This law generalizes the Darwinian principle of natural selection, the maximum entropy production principle [37] (entropy can also be described as information), and Kauffman's tentative fourth law of thermodynamics[3]. It is interesting that this law contains the second law of thermodynamics, as atoms interact, propagating information homogeneously. It also describes living organisms, where genetic information is propagated across generations. And it also describes cultural evolution, where information is propagated among individuals. Life is "far from thermodynamic equilibrium" because it constrains [32] the (more simple) information propagation at the thermodynamic scale, i.e. the increase of entropy, exploiting structures to propagate (or maintain) the (more complex) information at the biological scale.

In relation with the law of information transformation, as information requires agents to perceive it, information will be potentially transformed. This source of novelty will allow for the "blind" exploration of better ways of propagating information, according to the agents perceiving it and their environments.

Extending the previous example, if errors in transcription made by RNAP are beneficial for its propagation (which entails the propagation of the cell producing RNAP), cells with such novel proteins will have better chances of survival than their "cousins" without transcription errors.

The propagation of information can be classified as follows:

Autonomous. Information propagates by itself. Strictly speaking, this is not possible, since at least some information is determined by the environment. However, if more information is produced by itself than by its environment, we can call this autonomous propagation (See Section 5).

Symbiotic. Different information cooperates, helping to propagate each other.

Parasitic. Information exploits other information for its own propagation.

Altruistic. Information promotes the propagation of other information at the cost of its own propagation.

---

[3] "The workspace of the biosphere expands, on average, as fast as it can in this coconstructing biosphere" [32, p. 209]

4.3 Law of Requisite Complexity

Taking into account the law of information transformation, transformed information can increase, decrease, or maintain its previous complexity, i.e. amount [46]. However, *more complex information will require more complex agents to perceive, act on, and propagate it.* This law generalizes the cybernetic law of requisite variety [4]. Note that simple agents can perceive and interact with *part* of complex information, but they cannot (by themselves) propagate it. An agent cannot perceive (and thus contain) information more complex than itself. For simple agents, information that is complex for us will be simple as well. As stated above, different agents can perceive the same information in different ways, giving it different meanings.

The so called "arrow of complexity" in evolution [11] can be explained with this law. If we start with simple information, its transformation will produce by simple drift [39, 41] increases in the complexity of information, without any goal or purpose. This occurs simply because there is an open niche for information to become more complex as it varies. But this also promotes agents to become more complex to exploit novel (complex) information and propagate it. Evolution does not need to favour complexity in any way: information just propagates to every possible niche as fast as possible, and it seems that there is often an "adjacent possible" [32] niche of greater complexity.

For example, it can be said that a protein (as an agent) perceives some information via its binding sites, as it recognizes molecules that "fit" a site. More complex molecules will certainly need more complex binding sites. Whether complex molecules are better or worse is a different matter: some will be better, some will be worse. But for those which are better, the complexity of the proteins must match the complexity of the molecules perceived. If the binding site perceives only a part of the molecule, then this might be confused with other molecules which share the perceived part. Following the law of information transformation, there will be a variety of complexities of information. The law of requisite complexity just states that the increase in complexity of information is determined by the ability of agents to perceive, act on, and propagate more complex information.

Since more complex information will be able to produce more variety, the *speed* of the complexity increase will escalate together with the complexity of the information.

4.4 Law of Information Criticality

Transforming and propagating information will tend to a critical balance between its stability and its variability. Propagating information maintains itself as much as possible, but transforming information varies it as much as possible. This struggle leads to a critical balance analogous to the "edge of chaos" [36, 31], self-organized criticality [8, 1], and the "complexity from noise" principle [6]. The homeostasis of living systems can also be seen as the self-regulation of information criticality.

This law can generalize Kauffman's four candidate laws for the coconstruction of a biosphere [32, Ch. 8]. Their relationship with this framework demands further discussion, which is out of the scope of this paper.

A well known example can be seen with cellular automata [36] and random Boolean networks [31, 21, 23]: stable (ordered) dynamics limit considerably or do not allow change of states so information cannot propagate, while variable (chaotic) dynamics change the states too much, losing information. Following the law of information propagation, information will tend to a critical state between stability and variability to maximize its propagation: if it is too stable, it will not propagate, and if it is too variable, it will be transformed. In other words, "critical" information will be able to propagate better than stable or variable one, i.e. as fast as possible (cf. law of information propagation).

4.5 Law of Information Organization

Information produces constraints that regulate information production. These constraints can be seen as *organization* [32]. In other words, evolving information will be organized (by transformation and propagation) to regulate information production. According to the law of information criticality, this organization will lie at a critical area between stability and variability. And following the law of information propagation, the organization of information will enable it to propagate as fast as possible.

This law can also be seen as information having a certain *control* over its environment, since the organization of information will help it withstand perturbations. It has been shown [33, 47, 34] that using this idea as a fitness function can lead to the evolution of robust and adaptive agents, namely maximizing the mutual information between sensors and environment.

A clear example of information producing its own organization can be seen with living systems, which are discussed in Section 5.

4.6 Law of Information Self-organization

Information tends to its preferred, most probable state. This is actually a tautology, since observers determine probabilities after observing tendencies of information dynamics. Still, this tautology can be useful to describe and understand phenomena. This law lies at the heart of probability theory and dynamical systems theory [5]. The dynamics of a system tend to a subset of its state space, i.e. attractors, depending on its history. This simple fact reduces the possibility space of information, i.e. a system will tend towards a small subset of all possible states. If we describe attractors as "organized", then we can describe the dynamics of information in terms of self-organization [25].

Pattern formation can be described as information self-organizing, and related to the law of information propagation. Information will self-organize in "fit" patterns that are the most probable (defined *a posteriori*).

Understanding different ways in which self-organization is achieved by transforming information can help us understand better natural phenomena [24] and design artificial systems [22]. For example, random Boolean networks can be said to self-organize towards their attractors [23].

4.7 Law of Information Potentiality

An agent can give different potential meanings to information. This implies that the same information can have different meanings. Moreover, meaning— while being information—can be independent of the information carrying it, i.e. depend only on the agent observing it. Thus, different information can have the same potential meaning. The precise meaning of information will be given by an agent observing it within a specific context.

The potentiality of information allows the effective communication between agents. Different information has to be able to acquire the same meaning (homonymy), while the same information has to be able to acquire different meanings (polysemy) [44]. The relationship between the laws of information and communication is clear, but beyond the scope of this paper.

The law of information potentiality is related to a passive information transformation, i.e. a change in the agent observing information.

In spite of information potentiality, not all meanings will be suitable for all information. In other words, pure subjectivism cannot dictate meanings of information. By the law of information propagation, some meanings will be more suitable than others and will propagate. The suitability of meanings will be determined by their use and context [59]. However, there is always a certain freedom to subjectively transform information.

For example, a photon can be observed as a particle, as a wave, or as a particle-wave. The suitability of each given meaning is determined by the context in which the photon is described/observed.

4.8 Law of Information Perception

The meaning of information is unique for an agent perceiving it in unique, always changing open contexts. If meaning of information is determined by the use an agent makes of it, which is embedded in an open environment, we can go to such a level of detail that the meaning will be unique. Certainly, agents make generalizations and abstractions of perceptions in order to be able to respond to novel information. Still, the precise situation and context will never be repeated. This makes perceived information unique. The implication of this is that the response to any given information might be "unexpected", i.e. novelty can arise. Moreover, the meaning of information can be to a certain extent *arbitrary*. This is related with the law of information transformation, as the uniqueness of meaning allows the same information perceived differently by the same or different agents to be statically transformed.

This law is a generalization of the first law of human perception: "whatever is perceived can be perceived only from a uniquely situated place in the overall structure of points of view" [29, p. xxiv] (cited in [44, p. 250]). We can describe agents perceiving information as filtering it. An advantage of humans and other agents is that we can *choose* which filter to use to perceive. The suggestion is not that "unpleasant" information should be solipsistically ignored, but that information can be potentially actively transformed.

For example, T lymphocytes in an immune system can perceive foreign agents and attack them. Even when the response will be similar for similar foreign agents, each perception will be unique, a situation that always leaves space for novelty.

Scales of perception

Different information is perceived at different scales of observation [9]. As the scale tends to zero, then the information tends to infinite. For lower scales, more information and details are perceived. The uniqueness of information perception dominates at these very low (spatial and temporal) scales. However, as generalizations are made, information is "compressed", i.e. only relevant aspects of information are perceived[4]. At higher scales, more abstractions and generalizations are made, i.e. less information is perceived. When the scale tends to infinite, the information tends to zero. In other words, no information is needed to describe all of the universe, because all the information is already there. This most abstract understanding of the world is in line with the "highest view" of Vajrayana Buddhism [45]. Implications at this level of description cannot be right or wrong, because there is no context. Everything is contained, but no information is needed to describe it, since it is already there. This "maximum" understanding is also described as vacuity, which leads to bliss [45, p. 42].

Following the law of information criticality, agents will tend to a balance where the perceived information is minimal but maximally predictive [51] (at a particular scale): few information is cheaper, but more information in general entails a more precise predictability. The law of requisite complexity applies at particular scales, since a change of scale will imply a change of complexity of information [9].

5 On the Notion of Life

There is no agreed notion of life, which reflects the difficulty of defining the concept. Still, many researchers have put forward properties that characterize important aspects of life. *Autopoiesis* is perhaps the most salient one, which notes that living systems are self-producing [55, 38]. Still, it has been argued that autopoiesis is a necessary but not sufficient property for life [50]. The relevance of autonomy [10, 42, 35] and individuality [40, 35] for life have also been highlighted .

These approaches are not unproblematic, since no living system is completely autonomous. This follows from the fact that all living systems are open. For

---

[4]The relevance is determined by the context, i.e. different aspects will be relevant for different contexts.

example, we have some degree of autonomy, but we are still dependent on food, water, oxygen, sunlight, bacteria living in our gut, etc. This does not mean that we should abandon the notion of autonomy in life. However, we need to abandon the sharp distinction between life and non-life [11, 35], as different degrees of autonomy escalate *gradually*, from the systems we considered as non-living to the ones we consider as living. In other words, life has to be a fuzzy concept.

Under the present framework, living and non-living systems are information. Rather than a yes/no definition, we can speak about a *"life ratio"*:

Notion 4 *The ratio of living information is the information produced by itself over the information produced by its environment.*

Being more specific—since all systems also receive information—a system with a high life ratio produces more (first order) information about itself than the one it receives from its environment. Following the law of information organization, this also implies that living information produces more of its own constraints (organization) to regulate itself than the ones produced by its environment, and thus it has a greater autonomy. All information will have constraints from other (environmental) information, but we can measure (as second-order information) the proportion of internal over external constraints to obtain the life ratio. If this is greater than one, then the information regulates by itself more than the proportion that is regulated by external information. In the opposite case, the life ratio would be less than one.

Following the law of information propagation, evolution will tend to information with higher life ratios, simply because this can propagate better, as it has more "control" and autonomy over its environment. When information depends more on its environment for its propagation, it has a higher probability of being transformed as it interacts with its environment.

Note that the life ratio depends on spatial and temporal scales at which information is perceived. For example, for some microorganisms observed at a scale of years , the life ratio would be less than one, but if observed at a scale of seconds, the life ration would be greater than one.

Certainly, some artificial systems would be considered as living under this notion. However, we can make a distinction between living systems embodied in or composed by biological cells [16], i.e. life as we know it, and the rest, i.e. life as it could be. The latter ones are precisely those explored by artificial life.

6   On the Notion of Cognition

Cognition is certainly related with life [53]. The term has taken different meanings in different contexts, but all of them can be generalized into a common notion [20]. Cognition comes from the Latin *cognoscere*, which means "get to know". Like this,

Notion 5 *A system is cognitive if it knows something [20, p.135].*

From Notion 2, all agents are cognitive, since they "*know*" how to act on their environment, giving (first order) *meaning* to their environmental information. Thus, there is no boundary between non-cognitive and cognitive systems. Throughout evolution, however, there has been a *gradual* increase in the complexity of cognition [20]. This is because all agents can be described as possessing some form of cognition, i.e. "knowledge" about the (first-order) information they perceive[5].

Following the law of requisite complexity, evolution leads to more complex agents, to be able to cope with the complexity of their environment. This is precisely what triggers the (second-order) increase in the complexity of cognition we observe.

Certainly, there are different types of cognition[6]. We can say that a rock "knows" about gravity because it perceives its information, which has an effect on it, but it cannot *react* to this information. Throughout evolution, information capable of maintaining its integrity has prevailed over that which was not. *Robust* information is that which can resist perturbations to maintain its integrity. The ability to react to face perturbations to maintain information makes information *adaptive*, increasing its probability of maintenance. When this reaction is made before it occurs, the information is *anticipative*[7]. As information becomes more complex (even if only by information transformation), the mechanisms for maintaining this information also become more complex, as stated by the law of requisite complexity. This has led gradually to the advanced cognition that animals and machines posses.

7   Future Work

The ideas presented here still need to be explored and elaborated further. One way of doing this would be with a simulation-based method. Being inspired by ϵ-machines [51, 26], one could start with "simple" agents that are able to perceive and produce information, but cannot control their own production. These would be let to evolve, measuring if complexity increases as they evolve. The hypothesis is that complexity would increase (under which conditions still remains to be seen), to a point where "ϵ-agents" will be able to produce themselves depending more on their own information than that of the environment. This would be similar to the evolution in Tierra [48] or Avida [3] systems, only that self-replication would not be inbuilt. The tentative laws of information presented in Section 4 would be better defined if such a system was studied.

One important aspect that remains to be studied is the representation of

---

[5] One could argue that, since agency (and thus cognition) is already assumed in all agents, this approach is not explanatory. But I am not trying to explain the "origins" of agency, since I assume it to be there from the start. I believe that we can only study the evolution and complexification of agency and cognition, not their "origins".

[6] For example, human, animal, plant, bacterial, immune, biological, adaptive, systemic, and artificial [20].

[7] For a more detailed treatment on robustness, adaptation, and anticipation, see [22]

thermodynamics in terms of information. This is because the ability to perform thermodynamic work is a characteristic property of biological systems [32]. This work can be used to generate the organization necessary to sustain life (cf. law of information organization). It is difficult to describe life in terms of thermodynamics, since it entails new characteristic properties not present in thermodynamic systems. But if we see the latter ones as information, it will be easier to describe how life—also described as information—evolves from them, as information propagates itself at different scales.

A potential application of this framework would be in economy, considering capital, goods, and resources as information (a non-conserved quantity) [18]. A similar benefit (of non-conservation) could be given in game theory: if the payoff of games is given in terms of information (not necessarily conserved), nonzero sum games could be easier to grasp than if the payoff is given in material (conserved) goods.

It becomes clear that information (object), the agent perceiving it (subject) and the meaning-making or transformation of information (action) are deeply interrelated. They are part of the same totality, since one cannot exist without the others. This is also in line with Buddhist philosophy. The implications of an informational description of the world for philosophy have also to be addressed, since some schools have focussed on partial aspects of the object-subject-action trichotomy. Another potential application of the laws of information would be in ethics, where value can be described accordingly to the present framework.

8 Conclusions

This paper introduced general ideas that require further development, extension and grounding in particular disciplines. Still, a first step is always necessary, and hopefully feedback from the community will guide the following steps of this line of research.

Different metaphors for describing the world can be seen as different languages: they can refer to the same objects without changing them. And each can be more suitable for a particular context. For example, English has several advantages for fast learning, German for philosophy, Spanish for narrative, and Russian for poetry. In other words, there is no "best" language outside a particular context. In a similar way, I am not suggesting that describing the world as information is more suitable than physics to describe physical phenomena, or better than chemistry to describe chemical phenomena. It would be redundant to describe particles as information if we are studying only particles. The suggested approach is meant only for the cases when the physical approach is not sufficient, i.e. across scales, constituting an alternative worth exploring to describe evolution.

It seems easier to describe matter and energy in terms of information than vice versa. Moreover, information could be used as a common language across scientific disciplines [56].

112

Acknowledgements

I should like to thank Irun Cohen, Inman Harvey, Francis Heylighen, David Krakauer, Antonio del Río, Marko Rodriguez, David Rosenblueth, Stanley Salthe, Mikhail Prokopenko, Clément Vidal, and Héctor Zenil for their useful comments and suggestions.

Bibliography

[1] ADAMI, Christoph, "Self-organized criticality in living systems", *Phys. Lett. A* **203** (1995), 29–32.

[2] ADAMI, Christoph, "What is complexity?", *Bioessays* **24**, 12 (December 2002), 1085–1094.

[3] ADAMI, Chris, and C. Titus BROWN, "Evolutionary learning in the 2d artificial life system "Avida"", *Proc. Artificial Life IV* (R. BROOKS AND P. MAES eds.), MIT Press (1994), 377–381.

[4] ASHBY, W. Ross, *An Introduction to Cybernetics*, Chapman & Hall London (1956).

[5] ASHBY, W. Ross, "Principles of the self-organizing system", *Principles of Self-Organization* (Oxford,) (H. V. FOERSTER AND G. W. ZOPF, JR. eds.), Pergamon (1962), 255–278.

[6] ATLAN, H., "On a formal definition of organization", *J Theor Biol* **45**, 2 (June 1974), 295–304.

[7] ATLAN, Henri, and Irun R. COHEN, "Immune information, self-organization and meaning", *Int. Immunol.* **10**, 6 (1998), 711—717.

[8] BAK, Per, Chao TANG, and Kurt WIESENFELD, "Self-organized criticality: An explanation of the 1/f noise", *Phys. Rev. Lett.* **59**, 4 (July 1987), 381–384.

[9] BAR-YAM, Y., "Multiscale variety in complex systems", *Complexity* **9**, 4 (2004), 37–45.

[10] BARANDARIAN, Xabier, "Behavioral adaptive autonomy. a milestone in the ALife route to AI?", *Artificial Life IX Proceedings of the Ninth International Conference on the Simulation and Synthesis of Living Systems* (J. POLLACK, M. BEDAU, P. HUSBANDS, T. IKEGAMI, AND R. A. WATSON eds.), MIT Press (2004), 514–521.

[11] BEDAU, Mark A., "Four puzzles about life", *Artificial Life* **4** (1998), 125–140.

[12] COHEN, Irun R., *Tending Adam's Garden: Evolving the Cognitive Immune Self*, Academic Press London (2000).

[13] COHEN, Irun R., "Informational landscapes in art, science, and evolution", *Bulletin of Mathematical Biology* **68**, 5 (July 2006), 1213–1229.

[14] COVER, Thomas M., and Joy A. THOMAS, *Elements of Information Theory*, Wiley-Interscience (July 2006).

[15] DARWIN, Charles, *The Origin of Species*, Wordsworth (1998).

[16] DE DUVE, Christian, *Live Evolving: Molecules, Mind, and Meaning*, Oxford University Press (2003).

[17] DENNETT, Daniel, *Darwin's Dangerous Idea*, Simon & Schuster (1995).

[18] FARMER, J. Doyne, and Neda ZAMANI, "Mechanical vs. informational components of price impact", *EPJ B* **55**, 2 (2007), 189–200.

[19] FLORIDI, Luciano ed., *The Blackwell Guide to Philosophy of Computing and Information*, Blackwell (2003).

[20] GERSHENSON, Carlos, "Cognitive paradigms: Which one is the best?", *Cognitive Systems Research* **5**, 2 (June 2004), 135–156.

[21] GERSHENSON, Carlos, "Introduction to random Boolean networks", *Workshop and Tutorial Proceedings, Ninth International Conference on the Simulation and Synthesis of Living Systems (ALife IX)* (Boston, MA,) (M. BEDAU, P. HUSBANDS, T. HUTTON, S. KUMAR, AND H. SUZUKI eds.), (2004), 160–173.

[22] GERSHENSON, Carlos, *Design and Control of Self-organizing Systems*, CopIt Arxives Mexico (2007), http://tinyurl.com/DCSOS2007.

[23] GERSHENSON, Carlos, "Guiding the self-organization of random Boolean networks", *Theory in Biosciences* (In Press).

[24] GERSHENSON, Carlos, "The sigma profile: A formal tool to study organization and its evolution at multiple scales", *Complexity* (In Press).

[25] GERSHENSON, Carlos, and Francis HEYLIGHEN, "When can we call a system self-organizing?", *Advances in Artificial Life, 7th European Conference, ECAL 2003 LNAI 2801* (Berlin,) (W. BANZHAF, T. CHRISTALLER, P. DITTRICH, J. T. KIM, AND J. ZIEGLER eds.), Springer (2003), 606–614.

[26] GÖRNERUP, Olof, and James P. CRUTCHFIELD, "Hierarchical self-organization in the finitary process soup", *Artificial Life* **In Press** (2008), Special Issue on the Evolution of Complexity.

[27] HELBING, Dirk, and Tamás VICSEK, "Optimal self-organization", *New Journal of Physics* **1** (1999), 13.1–13.17.

[28] HEYLIGHEN, Francis, and Cliff JOSLYN, "Cybernetics and second order cybernetics", *Encyclopedia of Physical Science and Technology*, (R. A. MEYERS ed.) 3rd ed. vol. 4. Academic Press New York (2001), pp. 155–170.

[29] HOLQUIST, M., "Introduction", *Art and Answerability*, (M. M. BAKHTIN ed.). University of Texas Press Austin (1990).

[30] HOPFIELD, J. J., "Physics, computation, and why biology looks so different", *Journal of Theoretical Biology* **171** (1994), 53–60.

[31] KAUFFMAN, S. A., *The Origins of Order*, Oxford University Press (1993).

[32] KAUFFMAN, Stuart A., *Investigations*, Oxford University Press (2000).

[33] KLYUBIN, Alexander S., Daniel POLANI, and Chrystopher L. NEHANIV, "Organization of the information flow in the perception-action loop of evolved agents perception-action loop of evolved agents", *Proceedings of 2004 NASA/DoD Conference on Evolvable Hardware* (R. S. ZEBULUM, D. GWALTNEY, G. HORNBY, D. KEYMEULEN, J. LOHN, AND A. STOICA eds.), IEEE Computer Society (2004), 177–180.

[34] KLYUBIN, Alexander S., Daniel POLANI, and Chrystopher L. NEHANIV, "Representations of space and time in the maximization of information flow in the perception-action loop", *Neural Computation* **19** (2007), 2387–2432.

[35] KRAKAUER, D. C., and P. M. A. ZANOTTO, "Viral individuality and limitations of the life concept", *Protocells: Bridging Nonliving and Living Matter*, (S. RASMUSSEN, M. A. BEDAU, L. CHEN, D. DEAMER, D. C. KRAKAUER, N. PACKARD, AND D. P. STADLER eds.). MIT Press (2007).

[36] LANGTON, Christpher, "Computation at the edge of chaos: Phase transitions and emergent computation", *Physica D* **42** (1990), 12–37.

[37] MARTYUSHEV, L. M., and V. D. SELEZNEV, "Maximum entropy production principle in physics, chemistry and biology", *Physics Reports* **426**, 1 (April 2006), 1–45.

[38] MCMULLIN, Barry, "30 years of computational autopoiesis: A review", *Artificial Life* **10**, 3 (Summer 2004), 277–295.

[39] MCSHEA, Daniel W., "Metazoan complexity and evolution: Is there a trend?", *Evolution* **50** (1996), 477–492.

[40] MICHOD, Richard E., *Darwinian Dynamics: Evolutionary Transitions in Fitness and Individuality*, Princeton University Press Princeton, NJ (2000).

[41] MICONI, T, "Evolution and complexity: the double-edged sword", *Artificial Life* **14**, 3 (Summer 2008), 325–344, Special Issue on the Evolution of Complexity.

[42] MORENO, Alvaro, and Kepa RUIZ-MIRAZO, "The maintenance and open-ended growth of complexity in nature: information as a decoupling mechanism in the origins of life", *Reframing Complexity: Perspectives from the North and South*, (F. CAPRA, A. JUARRERO, P. SOTOLONGO, AND J. VAN UDEN eds.). ISCE Publishing (2006).

[43] MOROWITZ, Harold, and D. Eric SMITH, "Energy flow and the organization of life", *Tech. Rep. no. 06-08-029*, Santa Fe Institute, (2006).

[44] NEUMAN, Yair, *Reviving the Living: Meaning Making in Living Systems* vol. 6 of *Studies in Multidisciplinarity*, Elsevier Amsterdam (2008).

[45] NYDAHL, Ole, *The Way Things Are: A living Approach to Buddhism for today's world.*, O Books (2008).

[46] PROKOPENKO, Mikhail, Fabio BOSCHETTI, and Alex J. RYAN, "An information-theoretic primer on complexity, self-organisation and emergence", *Complexity* **15**, 1 (2009), 11–28.

[47] PROKOPENKO, M., V. GERASIMOV, and I. TANEV, "Evolving spatiotemporal coordination in a modular robotic system", *From Animals to Animats 9: 9th International Conference on the Simulation of Adaptive Behavior (SAB 2006)* (S. NOLFI, G. BALDASSARRE, R. CALABRETTA, J. C. T. HALLAM, D. MAROCCO, J.-A. MEYER, O. MIGLINO, AND D. PARISI eds.), vol. 4095 of *Lecture Notes in Computer Science*, Springer (2006), 558–569.

[48] RAY, T. S., "An approach to the synthesis of life", *Artificial Life II*, (C. LANGTON, C. TAYLOR, J. D. FARMER, AND S. RASMUSSEN eds.) vol. XI of *Santa Fe Institute Studies in the Sciences of Complexity*. Addison-Wesley Redwood City, CA (1991), pp. 371–408.

[49] ROEDERER, Juan G., *Information and its Role in Nature*, Springer-Verlag Heidelberg (May 2005).

[50] RUIZ-MIRAZO, Kepa, and Alvaro MORENO, "Basic autonomy as a fundamnental step in the synthesis of life", *Artificial Life* **10**, 3 (Summer 2004), 235–259.

[51] SHALIZI, Cosma R., *Causal Architecture, Complexity and Self-Organization in Time Series and Cellular Automata*, PhD thesis University of Wisconsin at Madison (2001).

[52] SHANNON, C. E., "A mathematical theory of communication", *Bell System Technical Journal* **27** (July and October 1948), 379–423 and 623–656.

[53] STEWART, John, "Cognition = life : Implications for higher-level cognition", *Behavioural processes* **35** (1995), 311–326.

[54] UMPLEBY, Stuart, "Physical relationships among matter, energy and information", *Cybernetics and Systems 2004* (Vienna,) (R. TRAPPL ed.), vol. 1, Austrian Society for Cybernetic Studies, (2004), 124–6.

[55] VARELA, Francisco J., Humberto R. MATURANA, and R. URIBE., "Autopoiesis: The organization of living systems, its characterization and a model", *BioSystems* **5** (1974), 187–196.

[56] VON BAEYER, Hans Christian, *Information: The New Language of Science*, Harvard University Press Cambridge, MA (2004).

[57] VON UEXKÜLL, Jakob, "A stroll through the worlds of animals and men", *Instinctive Behavior: The Development of a Modern Concept*, (C. H. SCHILLER ed.). International Universities Press New York (1957), pp. 5–80.

[58] WHEELER, John Archibald, "Information, physics, quantum: the search for links", *Complexity, Entropy, and the Physics of Information*, (W. H. ZUREK ed.) vol. VIII of *Santa Fe Institute Studies in the Sciences of Complexity*. Perseus Books Reading, MA (1990).

[59] WITTGENSTEIN, Ludwig, *Philosophical Investigations* 3rd ed., Prentice Hall (1999).

[60] WOLFRAM, Stephen, *A New Kind of Sciene*, Wolfram Media (2002).

Chapter 11

Inferring Diversity: Life After Shannon

Adom Giffin
Department of Physics
University at Albany–SUNY
Albany, NY 12222,USA

The diversity of a community that cannot be fully counted must be inferred. The two preeminent inference methods are the MaxEnt method, which uses information in the form of constraints and Bayes' rule which uses information in the form of data. It has been shown that these two methods are special cases of the method of Maximum (relative) Entropy (ME). We demonstrate how this method can be used as a measure of diversity that not only reproduces the features of Shannon's index but exceeds them by allowing more types of information to be included in the inference. A specific example is solved in detail. Additionally, the entropy that is found is the same form as the thermodynamic entropy.

1   Introduction

Diversity is a concept that is used in many fields to describe the variability of different entities in a group. In ecology, the Shannon entropy [1] and Simpson's index [2] are the predominate measures of diversity. In this paper we focus on the Shannon entropy for two reasons: First, it has been shown that Simpson's index is an approximation of Shannon's [3]. Second, Shannon's entropy is closely tied to many other areas of research, such as information theory and physics.

It is often the case that the species in a community cannot be fully counted. In this case, when one has incomplete information, one must rely on methods of inference. The two preeminent inference methods are the MaxEnt [4] method, which has evolved to a more general method, the method of Maximum (relative) Entropy (ME) [5, 6, 7] and Bayes' rule. The choice between the two methods has traditionally been dictated by the nature of the information being processed (either constraints or observed data). However, it has been shown that one can accommodate both types of information in one method, ME [8]. The purpose of this paper is to demonstrate how the ME method can be used as a measure of diversity that is able to include more information that Shannon's measure allows.

Traditionally when confronted with a community whose count is incomplete, the frequency of the species that are counted are used to calculate the diversity. The frequency is used because it represents an estimate of the probability of finding a particular species in the community. However, the frequency is not equivalent to the probability [9] and as such is a poor estimate. Fortunately, there are much better methods for estimating or inferring the probability such as MaxEnt and Bayes. Even more fortunate is that the new ME method can reproduce every aspect of Bayesian and MaxEnt inference *and* tackle problems that the two methods alone could not address.

We start by showing a general example of the ME method by inferring a probability with two different forms of information: expected values[1] and data, *simultaneously*. The solution resembles Bayes' Rule. In fact, if there are no moment constraints then the method produces Bayes rule *exactly*. If there is no data, then the MaxEnt solution is produced.

Finally we solve a toy ecological problem and discuss the diversity calculated by using Shannon's entropy and the diversity calculated by the ME method. This illustrates the many advantages to using the ME method.

2 Simultaneous updating

Our first concern when using the ME method to update from a prior to a posterior distribution[2] is to define the space in which the search for the posterior will be conducted. We wish to infer something about the values of one or several quantities, $\theta \in \Theta$, on the basis of three pieces of information: prior information about θ (the prior), the known relationship between x and θ (the model), and the observed values of the data $x \in \mathcal{X}$. Since we are concerned with both x and θ, the relevant space is neither $\mathcal{X}$ nor Θ but the product $\mathcal{X} \times \Theta$ and our attention must be focused on the joint distribution $P(x, \theta)$. The selected joint posterior $P_{\text{new}}(x, \theta)$ is that which maximizes the entropy,

$$
S[P, P_{\text{old}}] = - \int dx d\theta \; P(x, \theta) \log \frac{P(x, \theta)}{P_{\text{old}}(x, \theta)} \; , \tag{1}
$$

---

[1]For simplicity we will refer to these expected values as *moments* although they can be considerably more general.

[2]In Bayesian inference, it is assumed that one always has a prior probability based on some prior information. When new information is attained, the old probility (the prior) is *updated* to a new probability (the posterior). If one has no prior information, then one uses an *ignorant* prior [10].

subject to the appropriate constraints. $P_{old}(x, \theta)$ contains our prior information which we call the *joint prior*. To be explicit,

$$P_{old}(x, \theta) = P_{old}(\theta)P_{old}(x|\theta) , \qquad (2)$$

where $P_{old}(\theta)$ is the traditional Bayesian prior and $P_{old}(x|\theta)$ is the likelihood. It is important to note that they *both* contain prior information. The Bayesian prior is defined as containing prior information. However, the likelihood is not traditionally thought of in terms of prior information. Of course it is reasonable to see it as such because the likelihood represents the model (the relationship between θ and x) that has already been established. Thus we consider both pieces, the Bayesian prior and the likelihood to be *prior* information.

The new information is the *observed data*, x', which in the ME framework must be expressed in the form of a constraint on the allowed posteriors. The family of posteriors that reflects the fact that x is now known to be x' is such that

$$C_1 : P(x) = \int d\theta \, P(x, \theta) = \delta(x - x') . \qquad (3)$$

This amounts to an *infinite* number of constraints: there is one constraint on $P(x, \theta)$ for each value of the variable x and each constraint will require its own Lagrange multiplier $\lambda(x)$. Furthermore, we impose the usual normalization constraint,

$$\int dx d\theta \, P(x, \theta) = 1 , \qquad (4)$$

and include additional information about θ in the form of a constraint on the expected value of some function $f(\theta)$[3],

$$C_2 : \int dx d\theta \, P(x, \theta)f(\theta) = \langle f(\theta) \rangle = F . \qquad (5)$$

We emphasize that constraints imposed at the level of the prior need not be satisfied by the posterior. What we do here differs from the standard Bayesian practice in that we *require* the constraint to be satisfied by the posterior distribution.

Maximize (1) subject to the above constraints,

$$\delta \left\{ \begin{array}{c} S + \alpha \left[\int dx d\theta P(x, \theta) - 1 \right] \\ + \beta \left[\int dx d\theta P(x, \theta) f(\theta) - F \right] \\ + \int dx \lambda(x) \left[\int d\theta P(x, \theta) - \delta(x - x') \right] \end{array} \right\} = 0 , \qquad (6)$$

yields the joint posterior,

$$P_{new}(x, \theta) = P_{old}(x, \theta)\frac{e^{\lambda(x)+\beta f(\theta)}}{Z} , \qquad (7)$$

---

[3]Including an additional constraint in the form of $\int dx d\theta P(x, \theta)g(x) = \langle g \rangle = G$ could only be used when it does not contradict the data constraint (3). Therefore, it is redundant and the constraint would simply get absorbed when solving for $\lambda(x)$.

where Z is determined by using (4),

$$Z = e^{-\alpha+1} = \int dx d\theta e^{\lambda(x)+\beta f(\theta)} P_{\text{old}}(x,\theta) \tag{8}$$

and the Lagrange multipliers $\lambda(x)$ are determined by using (3)

$$e^{\lambda(x)} = \frac{Z}{\int d\theta e^{\beta f(\theta)} P_{\text{old}}(x,\theta)} \delta(x-x') . \tag{9}$$

The posterior now becomes

$$P_{\text{new}}(x,\theta) = P_{\text{old}}(x,\theta)\delta(x-x')\frac{e^{\beta f(\theta)}}{\zeta(x,\beta)} , \tag{10}$$

where $\zeta(x,\beta) = \int d\theta e^{\beta f(\theta)} P_{\text{old}}(x,\theta)$.

The Lagrange multiplier β is determined by first substituting the posterior into (5),

$$\int dx d\theta \left[P_{\text{old}}(x,\theta)\delta(x-x')\frac{e^{\beta f(\theta)}}{\zeta(x,\beta)} \right] f(\theta) = F . \tag{11}$$

Integrating over x yields,

$$\frac{\int d\theta e^{\beta f(\theta)} P_{\text{old}}(x',\theta) f(\theta)}{\zeta(x',\beta)} = F , \tag{12}$$

where $\zeta(x,\beta) \to \zeta(x',\beta) = \int d\theta e^{\beta f(\theta)} P_{\text{old}}(x',\theta)$. Now β can be determined by

$$\frac{\partial \ln \zeta(x',\beta)}{\partial \beta} = F . \tag{13}$$

The final step is to marginalize the posterior, $P_{\text{new}}(x,\theta)$ over x to get our updated probability,

$$P_{\text{new}}(\theta) = P_{\text{old}}(x',\theta)\frac{e^{\beta f(\theta)}}{\zeta(x',\beta)} \tag{14}$$

Additionally, this result can be rewritten using the product rule as

$$P_{\text{new}}(\theta) = P_{\text{old}}(\theta)P_{\text{old}}(x'|\theta)\frac{e^{\beta f(\theta)}}{\zeta'(x',\beta)} , \tag{15}$$

where $\zeta'(x',\beta) = \int d\theta e^{\beta f(\theta)} P_{\text{old}}(\theta) P_{\text{old}}(x'|\theta)$. The right side resembles Bayes theorem, where the term $P_{\text{old}}(x'|\theta)$ is the standard Bayesian likelihood and $P_{\text{old}}(\theta)$ is the prior. The exponential term is a *modification* to these two terms. Notice when $\beta = 0$ (no moment constraint) we recover Bayes' rule. For $\beta \neq 0$ Bayes' rule is modified by a "canonical" exponential factor.

It must be noted that MaxEnt has been traditionally used for obtaining a prior for use in Bayesian statistics. When this is the case, the updating is sequential. This is not the case here where both types of information are processed simultaneously. In the sequential updating case, the multiplier β is chosen so that the posterior P_{new} only satisfies C_2. In the simultaneous updating case the multiplier β is chosen so that the posterior P_{new} satisfies both C_1 and C_2 or $C_1 \wedge C_2$ [8].

3   Inference in Ecology

In the following sections we will discuss the traditional way diversity is measured and the way it is measured using ME. This will be done by examining a simple example and comparing the two methods. In addition, we will show how the ME method could include information that the traditional method cannot.

The general information for the example is as follows: There are k types of plants in a forest. A portion of the forest is examined and the amount of each species is counted where $m_1, m_2 \ldots m_k$ represents the counts of each species and n represents the total count so that $n = \sum_i^k m_i$. Additionally, we know from biological examination that one species, s_2 and another species, s_5 are codependent. Perhaps they need each others pollen in such supply that they cannot exist unless there are on the average, twice the number of s_2 as compared to s_5.

3.1   Traditional Diversity

We calculate the Shannon diversity by using Shannon's entropy as follows,

$$S_{Tradtional} = -\sum_i^k p_i \log p_i \,, \tag{16}$$

where $p_i = m_i/n$. The problem with using this method is not in the method itself but with the reason it is being used. If the purpose of using this method was to measure the diversity of the portion that was counted then the method is acceptable. However, if the purpose of the method is to estimate or infer the diversity of the *whole* forest, then it is a poor estimate. First, p_i is meant to represent the probability of finding the *ith* species in the forest. As previously stated, the frequency of the sample is not equivalent to the probability. In fact, it is the expected value of the frequency that is equivalent to the probability, $\langle F \rangle = p$ [9]. It would only make sense to use the frequency as an estimate of the probability when n is very large (i.e. $n \to \infty$) but this is not usually the case. Second, the diversity of two samples that have the same ratio of frequencies will be the same. Therefore this measure does not reflect the abundance of the species. This might be a desirable feature [3]. Third, there is no clear way to process the information about the codependence using Shannon's entropy.

3.2   ME Diversity

Here we intend to use a better method to estimate or infer p_i and that method is the ME method. The first task is to realize that the correct mathematical model for the probability of getting a particular species where the information that we have is the number of species counted is a multinomial distribution. The probability of finding k species in n counts which yields m_i instances for the ith

species is

$$P_{\text{old}}(m|p,n) = P_{\text{old}}(m_1 \ldots m_k|p_1 \ldots p_k, n) = \frac{n!}{m_1! \ldots m_k!} p_1^{m_1} \ldots p_k^{m_k} , \tag{17}$$

where $m = (m_1, \ldots, m_k)$ with $\sum_{i=1}^{k} m_i = n$, and $p = (p_1, \ldots, p_k)$ with $\sum_{i=1}^{k} p_i = 1$. The general problem is to infer the parameters p on the basis of information about the data, m'. Here we see the first advantage with using the ME diversity; we allow for fluctuations in our inference by looking at a distribution of $p's$ as opposed to claiming that we know the "true" p.

Additionally we can include information about the codependence by using the following general constraint,

$$\langle f(p) \rangle = F \quad \text{where} \quad f(p) = \sum_{i}^{k} f_i p_i , \tag{18}$$

where f_i is used to represent the codependence. For our example, on the average, we will find twice the number of s_2 as compared to s_5 thus, *on the average*, the probability of finding one of the species will be twice that of the other, $\langle p_2 \rangle = 2 \langle p_5 \rangle$. In this case, $f_2 = 1$, $f_5 = -2$ and $f_{i \neq (2,5)} = F = 0$.

Next we need to write the data (counts) as a constraint which in general is

$$P(m|n) = \delta_{mm'} , \tag{19}$$

where $m' = \{m'_1, \ldots, m'_k\}$. Finally we write the appropriate entropy to use,

$$S[P, P_{\text{old}}] = -\sum_m \int dp \, P(m, p|n) \log \frac{P(m, p|n)}{P_{\text{old}}(m, p|n)} , \tag{20}$$

where

$$\sum_m = \sum_{m_1 \ldots m_k = 0}^{n} \delta(\sum_{i=1}^{k} m_i - n) , \tag{21}$$

and

$$\int dp = \int dp_1 \ldots dp_k \, \delta \left(\sum_{i=1}^{k} p_i - 1 \right) , \tag{22}$$

and where $P_{\text{old}}(m, p|n) = P_{\text{old}}(p|n) P_{\text{old}}(m|p, n)$. The prior $P_{\text{old}}(p)$ is not important for our current purpose so for the sake of definiteness we can choose it flat for our example (there are most likely better choices for priors). We then maximize this entropy with respect to $P(m, p|n)$ subject to normalization and our constraints which after marginalizing over m' yields,

$$P(p) = P_{\text{old}}(m'|p, n) \frac{e^{\beta f(p)}}{\zeta} , \tag{23}$$

where

$$\zeta = \int dp \, e^{\beta f(p)} P_{\text{old}}(m'|p, n) \quad \text{and} \quad F = \frac{\partial \log \zeta}{\partial \beta} . \tag{24}$$

121

The probability distribution $P(p)$ has sometimes been criticized for being too strange. The idea of getting a probability of a probability may seem strange at first but makes absolute sense. We do not know the "true" distribution of species, p_i. Therefore it seems natural to express our knowledge with some uncertainty in the form of a distribution. Notice that if one has no information relating the species then $\beta = 0$.

Finally by substituting (23) into (20), and using our constraints (18) and (19) we introduce our new general measure for diversity,

$$S_{ME} = \log \zeta - \beta F . \tag{25}$$

4 Conclusions

Diversity is an important concept in many fields. In this paper we provided a toy example of how ME would be used as a measure of diversity that may simulate real world situations. By using the multinomial, we not only properly infer p so that fluctuations are represented, we get the additional bonus of having the abundance of the species represented in the measure. It is critical to note that our diversity, S_{ME} satisfies all of Pielou's axioms [11].

This of course could all be done with only using Bayes to infer p. However, by using the ME method we can include additional information allowing us to go beyond what Bayes' rule and MaxEnt methods alone could do. Therefore, we would like to emphasize that anything one can do with Bayesian or MaxEnt methods, one can now do with ME. Additionally, in ME one now has the ability to apply additional information that Bayesian or MaxEnt methods could not. Further, any work done with Bayesian techniques can be implemented into the ME method directly through the joint prior.

Although Shannon had discovered the entropy that bears his name quite independently of thermodynamic considerations, it nevertheless is directly proportional to the thermodynamic entropy. The realization that the ME diversity is of the exact same form as the thermodynamic entropy[4] is of no small consequence. All of the concepts that thermodynamics utilizes can now also be utilized in ecology, whether it be energy considerations or equilibrium conditions, etc.

To see a detailed method for calculating ζ, see [8], for a numeric example, see [12] and for an example of what do when one knows that there are species in the forest but simply have not been counted (perhaps they are rare), see [13].

Acknowledgements: We would like to acknowledge many valuable discussions with A. Caticha

---

[4]The thermodynamic entropy is actually, $S = \log \zeta + \beta F$. The fact that our entropy (25) has a $-\beta F$ is a reflection of our choice to add our Lagrange multipliers in (6) as opposed to subtracting them as is the case in thermodynamics. However, this is trivial because when one solves for β in (13) the sign will be accounted for. Thus, if the Lagrange multiplier was subtracted, the solution to (13) would be $-F$ and the entropy would have a $+\beta F$.

Bibliography

[1] C. E. Shannon, "A Mathematical Theory of Communication", *Bell System Technical Journal*, **27**, 379 (1948).

[2] E. H. Simpson, "Measurement of Diversity", *Nature*, **163**, 688 (1949).

[3] R. Guiasu and S. Guiasu, "Conditional and Weighted Measures of Ecological Diversity", *International J. of Uncertainty*, **11**, 283 (2002).

[4] E. T. Jaynes, Phys. Rev. **106**, 620 and **108**, 171 (1957); R. D. Rosenkrantz (ed.), *E. T. Jaynes: Papers on Probability, Statistics and Statistical Physics* (Reidel, Dordrecht, 1983); E. T. Jaynes, *Probability Theory: The Logic of Science* (Cambridge University Press, Cambridge, 2003).

[5] J. E. Shore and R. W. Johnson, IEEE Trans. Inf. Theory **IT-26**, 26 (1980); IEEE Trans. Inf. Theory **IT-27**, 26 (1981).

[6] J. Skilling, "The Axioms of Maximum Entropy", *Maximum-Entropy and Bayesian Methods in Science and Engineering*, G. J. Erickson and C. R. Smith (eds.) (Kluwer, Dordrecht, 1988).

[7] A. Caticha and A. Giffin, "Updating Probabilities", *Bayesian Inference and Maximum Entropy Methods in Science and Engineering*, ed. by Ali Mohammad-Djafari (ed.), AIP Conf. Proc. **872**, 31 (2006) (http://arxiv.org/abs/physics/0608185).

[8] A. Giffin and A. Caticha, "Updating Probabilities with Data and Moments", *Bayesian Inference and Maximum Entropy Methods in Science and Engineering*, ed. by Kevin Knuth, et all, AIP Conf. Proc. **954**, 74 (2007) (http://arxiv.org/abs/0708.1593).

[9] D. S. Sivia, *Data Analysis: A Bayesian Tutorial* (Oxford U. Press, 1996).

[10] A. Gelman, et al., *Bayesian Data Analysis, 2nd edition* (CRC Press, 2004).

[11] E. Pielou. *Ecological Diversity* (Wiley, New York 1975).

[12] A. Giffin, "Updating Probabilities with Data and Moments: An Econometric Example", presented at the *3rd Econophysics Colloquium*, Ancona, Italy, 2007 (http://arxiv.org/abs/0710.2912).

[13] A. Giffin, "Updating Probabilities with Data and Moments: A Complex Agent Based Example", presented at the *7th International Conference on Complex Systems*, Boston, 2007.

Chapter 12

Life at the "edge of chaos" in a genetic model

Y. Grondin, D. J. Raine
Centre for Interdisciplinary Science, Department of Physics and
Astronomy, University of Leicester, Leicester LE1 7RH, UK
iscience@le.ac.uk
and V. Norris
Assemblages Moléculaires: Modélisation et Imagerie SIMS, FRE
CNRS 2829, Faculté des Sciences et Techniques de Rouen, F76821
Mont Saint Aignan, France &
Epigenomics Project, genopole®, 93 rue Henri-Rochefort, 91000 Evry,
France

Genetic regulatory networks are often represented as Boolean networks and characterised by average parameters such as internal homogeneity (the probability that a node outputs a 1). Here we present a different formalism in which the nodes interact through positive and negative links with the state of the nodes determined by a single, general logical function. The main parameter of the system is in this case the proportion of negative links. At a critical value of the proportion of negative links the networks display a phase transition from a periodic to a chaotic regime The proportion of negative links in the bacterium Escherichia coli corresponds in our model to a range where the network behaves at the edge of chaos.

1 Introduction

The genetic regulatory network of a cell is a complex dynamical system yielding a wide diversity of living cells and organisms. Specific variations at any small or large scale in the level of expression of the genes, in their timing of action and in the architecture of the network are amongst the factors responsible for such diversity [1]. Boolean networks, a general form of cellular automata, have long been used to study the dynamical properties of such biological systems at both small and large scale [2,3]. For example, Boolean networks are used to represent the genetic regulatory networks inferred from mRNA data [4] and as models of cell differentiation [3,5].

Here, we use a variant of Boolean networks to model genetic regulation and to study properties related to architecture and dynamics. In the Boolean network introduced by Kauffman [3], also called the NK-model, the state of each of N nodes is represented by a Boolean variable (ON or OFF) where the output value is determined by a Boolean function, or transition function, that has for input the K connections to that node. The model is characterised by parameters such as the internal homogeneity, p, that is the probability that an output is ON [3,6], and the size of the stable core given by the number of stable nodes, that is the nodes that have a constant state independent of the initial state [7]. This model displays a complex dynamics. A major result of the NK-model [3] is the phase transition occurring at $K = 2$ between a crystallised phase $(K < 2)$ and a chaotic phase $(K > 2)$. For $K = 2$, the network is said to operate at the "edge of chaos" [3].

From the point of view of genetic regulation, the traditional representation of Boolean networks does not capture adequately the regulatory mechanism of the genes. We choose a different representation by assigning to the links either a positive or a negative regulatory effect in a way similar to the activating or inhibiting effect of regulatory genes on those that they regulate. Although this could be modelled in the framework of classical Boolean networks, our results differ in many aspects and, allow comparisons to real biological data.

We first introduce the genetic regulatory model and then show that a critical regime occurs for two ranges of the proportion of inhibitor links, μ . Further analysis of this critical regime shows that the network behaves at the edge of chaos. Though this transition is usually observed for variation of the internal homogeneity [6,8], our study shows that μ and p are different parameters. Furthermore, we find that the fraction of negative links in this critical regime corresponds to that found in the transcriptional network of the bacterium Escherichia coli, suggesting that it too behaves at the edge of chaos.

2 The Model

2.1 Architecture of the model

We consider directed networks where the agents, or nodes, represent the cellular machinery of gene regulation and the links represent the regulating influence of the agents on each other. The links, which are fixed, can either have an activating or inhibiting effect on the nodes to which they are connected. A network is represented by its adjacency matrix A, with elements a_{ij} given by 0 if there is no link from node j to node i; 1 if node j is connected and directed to node i and acts as an inducer on i; and -1 if node j is connected and directed to node i and acts as a repressor on i.

We denote by μ the proportion of repressors in the network, that is the number of negative links in the matrix A as a fraction of the number of non-zero links. The mean connectivity of the network is $k = <\Sigma\ aij>$.

2.2 Dynamics

As for Boolean networks, each agent of the network is characterised by its binary state. The configuration of the network at any one time is given by a vector $S(t)$, where the element $s_i(t)$ is the state of agent i at time t, such that $s_i(t) = 0$ if the agent is OFF and $s_i(t) = 1$ otherwise.

The dynamics of the network is provided by a simple rule in which the state of the nodes at a given time step depends only on the configuration of the network at the previous time. This rule states that a node is ON if the number of active positive incoming links to the node is greater than the number of active negative ones. Furthermore, only the nodes that are ON can exert their control over the other nodes: that is a node that is OFF does not exert any control on other nodes whatever its outgoing links may be. This translates to the following expression in which a node i is ON at $t + 1$ if

$$\sum_j a_{ij} s_j(t) > 0 \qquad (1)$$

and otherwise the node is OFF. A consequence is that if all of the nodes connecting a node i are OFF, this node does not receive signals; this node is OFF by default. The nodes that remain OFF in this way during the simulation belong to the inactive core of the network. The other nodes are part of the active core.

Although the activation function in equation (1) could be expressed as a Boolean function, this would neither be the most appropriate approach, nor the simplest. The function given in (1) is a more general function than the standard activation-inhibition functions, where a node is activated if there is at least an activator but no inhibitor and not activated otherwise. Indeed, expression (1) carries a simple symmetrical cooperative effect between activators and between inhibitors: the more activator the more likely a node will be activated and, conversely, with the more inhibitors the more likely a node will be inhibited. Finally, observations suggest that the binding of a regulator to specific DNA sequences may not be binary, that is designed to bind or not. On the contrary, there are apparently thresholds conditioned, for example, by the specificity of the sequence or the concentration of the regulator [13]. Nevertheless for simplicity in this paper we set the threshold to zero.

2.3 External input

We consider that the networks are not autonomous, that is they cannot exhibit a dynamics in the absence of a constraining environment. A subset of the nodes is chosen to receive an external input, or signal, which are subsequently considered as input nodes. There are several possibilities for the choice of those nodes: the nodes without incoming links, a subset of those, a subset of any nodes, etc. However, for simplicity, the subset of input nodes is chosen at random regardless of the connectivity of the nodes. The chosen nodes remain ON at any time regardless the value of equation (1), which provides a clamping effect on the network.

3 Result

In the following, we consider two different network architectures: a random network in which links are formed between pairs of nodes at random [10] and a power-law network [14]. These networks differ in many aspects such as in their degrees of clustering and diameter [10]. The networks presented below are constructed with a number of nodes and a mean connectivity of the order of magnitude that is observed in typical bacterial models, that is of the order of 1000 nodes and a small mean connectivity, in view of the sparseness of the regulatory networks in cellular organisms [15].

At the start of a simulation, all the nodes of the network are set to be OFF. A number of nodes, set to $I = 50$ in the following, is selected at random with equal probability, to receive an external input. The state of each node is then repeatedly updated until either an attractor is found, which occurs after a transient phase, or until a maximum number of set time steps, L, is reached.

3.1 Phases

For μ varying between 0 and 1, simulations show that for random networks with $N = 1500$ nodes and $k = 8.0$ the model displays three different behaviours characterised by the nature of the attractor. The network is (i) crystallised if the attractor is a fixed point, (ii) periodic if the attractor cycles over a length of time $< L$ and (iii) considered to be chaotic if no period is found, that is the network has a period $> L$. This relates to the fact that for large scale networks a period may not be reached in a reasonable computational time, despite the fact that in a finite deterministic model a period must exist. The three observed behaviours are characteristic of classical Boolean networks and cellular automata [3,6,16].

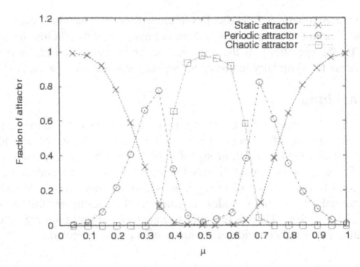

Figure 1. Fraction of a given attractor according to the fraction of negative links. The curves represent the fraction of static attractors (cross), periodic attractors (open circle) and chaotic attractors (open square) according to the proportion of negative links. Each point gives the fraction of network of a given attractor over 500 repeats. For each repeat, a random network of $N = 1500$, $k = 8.0$ is generated and a random subset of 50 nodes chosen as external inputs. Each simulation runs for a maximum of 105 time steps beyond which an attractor is considered chaotic.

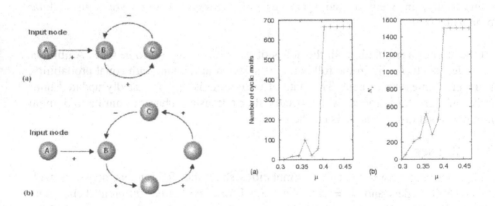

Figure 2 (left): Illustration of local and propagating structures. In both cases, the nodes A and C compose the neighbourhood of the node B. (a) The oscillation of the state of node a spreads locally to its neighbourhood whilst in (b) the oscillation spreads outside the neighbourhood.

Figure 3 (right): Transition in a single network for variation of the proportion of negative links. A network of 1500 nodes is initially constructed with μ = 0.3. The proportion of negative links is then increased and the number of nodes of variable states and the number of cyclic patterns formed from the nodes of variable state at the steady state is recorded. The variation of the number of cyclic motifs is given in (a) and the number of nodes of variable state in (b).

As shown in Fig. 1 for $L = 10^5$, the probability of finding a particular attractor depends on the value of μ. Thus, for $\mu < 0.27$ and $\mu > 0.78$, the network is likely to have a have a fixed point attractor while for $0.27 < \mu < 0.38$ and $0.65 < \mu < 0.78$ the network is more likely to reach a periodic attractor with a peak at $\mu \sim 0.35$ and $\mu \sim 0.70$. Finally, for $0.38 < \mu < 0.65$ the network is most likely to be in the chaotic regime with a symmetry in the distribution at $\mu \sim 0.53$.

Networks with a power-law distribution of the degree of connectivity show identical results (data not shown), which suggests that the architecture does not influence the behaviour of the model in this respect.

Further information is required to determine whether the periodic behaviour of the network has properties similar to that of cellular automata at the edge of chaos. In cellular automata, the probability for a node to be ON depends on a parameter λ, that is identical to the internal homogeneity of Boolean networks [16]. For λ close to 0 the system presents no activity after a very short number of steps. For λ around 0.2 some oscillatory states will persist either locally or propagating through the system in what are defined either as local or as propagating structures, respectively [16]. For λ around 0.3 those structures start to interact in complex patterns and when λ reaches 0.5 the system has become chaotic [16]. Here, a local structure is characterised by a periodic pattern confined to its neighbourhood (a node and the nodes it is directly connected to), while a propagating structure is a pattern that travels across neighbourhoods. In the network model, these structures are characterised by specific motifs. For example, the oscillatory state of node B generated by the structural motif in Fig.2(a) can spread only to its neighbouring node C, and similarly, the oscillatory state of node C can spread only to its neighbouring node B. The oscillatory pattern formed by the variation of the states of the nodes B and C is therefore local. Introducing a series of nodes between the nodes B and C, as shown in Fig. 2(b), allows the oscillatory states to propagate beyond its neighbourhood. The oscillatory pattern in the neighbourhood of node B is then a propagating structure.

To test whether the network behaves at the edge of chaos we need to look first at the number of local and propagating structures observed in a given network as the proportion of negative links increases [16]. This is equivalent to looking at the number of cyclic motifs in the part of the network formed by only the nodes of variable state. In addition, we need to check whether these structures are interacting with each other, that is, whether these structures can be connected by paths between the nodes of variable state. The number of cyclic motifs for a network of 1500 nodes and k = 8.0 is shown in Fig. 3(a) for a given network, and a given set of input nodes, as μ increases from 0.3 to 0.46. The number of cyclic motifs is measured by investigating the network made up of the nodes of variable state. First, we list all the nodes that are connected to a chosen starting node. Those nodes become starting nodes and the nodes connected to them are subsequently included in the list. This is repeated until no new node is added to the list. We then extract the nodes that appear at least twice in the list as they are either part of parallel pathways, as in a feed-forward loop for example [17], or they belong to a cycle, or both. Each of the selected nodes is then considered as the potential start of a cyclic pathway with the condition that once any such starting node has been identified as part of a cycle it cannot be part of any other cycles. This tends to under-estimate the number of cycles although it does not impair the result as shown below.

We also show in Fig. 3(b) the number of nodes of variable state forming the network. Thus for a small proportion of negative links ($\mu \sim 0.3$), where the network is likely to be crystallised, the number of cyclic motif is small (Fig. 4(a)). As the proportion of negative links increases to about $\mu \sim 0.39$, the number of cycles rises by one order of magnitude. Beyond this value where the network is likely to be chaotic (Fig. 1), the number of cyclic motifs varies as N, despite the under-estimation of the number of cyclic patterns. Note that all these structures are propagating: the probability of having a simple local structure such that of Fig. 2(a) is proportional to $k\,N^2$, hence close to zero in a network of small connectivity. Figure 3(b) shows also that the size of the network made of nodes of variable states varies similarly to the number of cyclic motifs. Finally, the measured number of independent components is less than 3, meaning that the propagating structures are interacting with each other. This demonstrates that for a range of μ, the network behaves at the edge of chaos.

4. Discussion

Classically, the behaviour of Boolean networks is affected by the bias introduced by the internal homogeneity parameter, p [3,6]. In the present model, the behaviour of the network is similarly affected by the proportion of negative links, μ. However, the parameters p and μ are noticeably different, this for the two following reasons. First, the value of p calculated from expression (1) for given values of μ does not equal μ. For example, in a structural motif where one incoming link is positive and another one is negative, that is $\mu = 0.5$, the probability for a node to be ON according to (1) is $p = 0.25$. Second, in the classical case, each value of p corresponds to a set of Boolean functions, whereas in our model each value of μ corresponds to a specific p calculated according to (1), but also to a Boolean function. Expression (1) determines the only possible Boolean function for a given value of μ and a given number of incoming links, ensuring that the logic of the transition functions is constant over the range of $\tilde{\mu}$. This has the modelling advantage that μ together with (1) provides a control parameter over the internal homogeneity of the system.

The rather different formalism adopted in the representation of the interactions between genes allows us to make comparisons with real genetic regulatory networks. For instance in RegulonDB, the transcriptional network of the bacterium Escherichia coli is described in term of the activating, inhibiting or dual function of regulators on the genes they regulate [18]. Considering the dual effect as a neutral one, the proportion of negative links to that of the total number of links, excluding those with a dual effect, gives $\mu = 0.4$. This corresponds in our model to a range at which the network is almost as likely to be periodic as to be chaotic, that is, it operates at the edge of chaos. This suggests that, similarly to the model, the transcriptional network of E. coli operates at the edge of chaos. Furthermore, the value of μ at which the networks are more likely to be periodic spans a small range, meaning that the value of μ for real systems may not be unique. However, because of the need for both robustness and adaptability we do expect a fine tuning of this value. Indeed, a small increase in μ and the networks may become too sensitive to perturbations, while conversely a small decrease in μ and the networks may become unresponsive to a change of vital signals. Note that such direct comparison would not be possible using the internal homogeneity as the key parameter.

Bibliography

[1] Adam S. Wilkins, The evolution of developmental pathways. (Sinauer Associates, 2002).

[2] D. Thieffry and D. Romero, Biosystems 50 (1), 49 (1999).

[3] S. A. Kauffman, The origins of order: Self-Organization and Selection in Evolution. (Oxford University Press, Oxford, 1993).

[4] S. Liang, S. Fuhrman, and R. Somogyi, Pac. Symp. Biocomput., 18 (1998); T. Akutsu, S. Miyano, and S. Kuhara, Bioinformatics 16 (8), 727 (2000); S. Martin, Z. Zhang, A. Martino et al., Bioinformatics (2007).

[5] S. Huang, J. Mol. Med. 77 (6), 469 (1999).

[6] G. Weisbuch and D. Stauffer, J. Physique 48 (1), 11 (1987).

[7] S. Bilke and F. Sjunnesson, Phys. Rev. E. 65, 016129 (2002).

[8] L. de Arcangelis and D. Stauffer, J. Physique 48 (11), 1881 (1987).

[9] R. Albert and A.-L. Barabási, Phys. Rev. Lett. 84 (24), 5660 (2000).

[10] R. Albert and A.-L. Barabasi, Rev. of Mod. Phys. 74 (1), 47 (2002).

[11] D. J. Watts and S. H. Strogatz, Nature 393 (6684), 440 (1998).

[12] B. Yuan, K. Chen, and B-H. Bang, cond-mat/0408391 (v1) (2004).

[13] U. Gerland, J. D. Moroz, and T. Hwa, Proc. Natl. Acad. Sci. U S A 99 (19), 12015 (2002).

[14] A.-L. Barabasi and R. Albert, Science 286 (5439), 509 (1999).

[15] N. Guelzim, S. Bottani, P. Bourgine et al., Nat. Genet. 31 (1), 60 (2002).

[16] C. G. Langton, Phys. D 2 (1-3), 120 (1986).

[17] R. Milo, S. Shen-Orr, S. Itzkovitz et al., Science 298 (5594), 824 (2002).

[18] H. Salgado, S. Gama-Castro, A. Martinez-Antonio et al., Nucleic Acids Res. 32 (Database issue), D303 (2004).

This work was performed using the University of Leicester Mathematical Modelling Centre's Supercomputer which was purchased through the EPSRC strategic equipment initiative.

Chapter 13

The Effects of Complexity on the Product-Service Shift

Ella-Mae Molloy[1], Dr. Carys Siemieniuch[2], Murray Sinclair[1]
[1]Dept. Human Sciences, [2]Dept. Electronic & Electrical Engineering
Loughborough University, Loughborough, UK
E.Molloy@lboro.ac.uk

There are many challenges facing businesses in the modern commercial climate. One challenge is the 'product-service shift', whereby organisations go from delivering a product to the provision of through-life availability of an instantiated capability. This will have implications for any organisation in the processes and procedures it employs, especially for knowledge and information management. Whilst it is true that there is a large technical content within the area of knowledge and information management, it is still, essentially, a human and organisational issue. This is the focus of the KIM (Knowledge and Information Management – through life) Grand Challenge.

This paper will introduce the KIM project and aims. Focus will be on a particular area of research within the project, concerned with decision-making, decision support and human aspects of such systems. Ongoing studies within this research are discussed, looking at the bid stage for major aerospace projects and at decision networks within flagship construction projects.

The implications of complexity, in its organisational guises of intrinsic and induced complexity will be discussed.

There is a discussion on how the limited information horizon contributes to (often unwanted) emergent behaviour as projects progress, and how decision-making systems may be configured to help cope with issues of complexity and the surprises such emergent behaviour can manifest.

1 Introduction

1.1 The KIM Grand Challenge

The Engineering and Physical Sciences Research Council (EPSRC) in the UK has funded five Grand Challenges. All involve academic and industrial partners and aim to improve the performance of UK businesses (specifically in manufacturing and construction). 'Knowledge and Information Management – Through Life' (KIM) is the biggest of these Grand Challenges, being funded at approximately £5 million (US $11 million).

KIM involves 11 academic institutions and number of industrial partners. Application and use in industry is a big focus of the output of this research. The overall aim of the project, as stated on the project website (www.kimproject.org), is:

> "The identification of approaches to information and knowledge management that may be applied to the through-life support of long lived, complex engineered products."

The work is split into four work packages and each of those into sub tasks. Each sub task has a team of people working on it from different universities. Loughborough University is involved with several of the sub tasks. However, the focus of the research reported in this paper is the third sub task within the third work package - 'Managing the Knowledge System Life Cycle'.

1.2 The Scope of Task 3.3

The task objective is to evaluate the potential of alternative methods of decision support in improving the quality of decision-making throughout the project lifecycle.

The team personnel are from the University of Salford, the University of Cambridge, the University of Reading and Loughborough University.

The scope of the research is to investigate the effect of a shift from product delivery to through life service support on decision-making and decision support in teams 'through life'. As noted on the project website, timescales may be '10, 20 or 30 years and beyond, during which time the 'information and knowledge' will be stored, accessed, used and received many times over in many different situations and contexts.'

The issues which arise from consideration of such a time frame can be summarised, as follows: beyond 30 years we will be designing (and servicing) products and systems whose requirements and uses not yet known, using materials and processes not yet invented or developed, using suppliers who will be very different to now, and who may have a shorter lifetime than the system itself, and all of this will be done, including maintaining system information and knowledge, with people who are not yet born and who will not be in post for more than a decade. (This has been termed as the 'looking forwards problem'.) There are also things to be learned 'looking backwards', i.e. how are legacy systems dealt with that were first designed and produced many years ago (e.g. the RAF Canberra photo reconnaissance aeroplane, designed in the 1950s and still flying today).

1.3 The Implications

The implications of a shift to 'through-life' service are not insignificant, not simply due to an extension to the traditional supply chain to cover service/support providers, but also because of the extended timescales and responsibilities to be considered.

Many of the customer's harder decisions about the future are delegated to the supplier (who can no longer pass risks on along the chain). The supplier, following Ashby's Law [Ashby 1956], must create and maintain a much more complex organisational structure to assess the issues arising. Immediate considerations are: (a) provision of the capability and its upgrades on time; (b) maintaining the availability of the capability after delivery for its lifetime; and (c) the provision of a problem resolution capability to support (a) and (b) above, by addressing unexpected happenings affecting the capability.

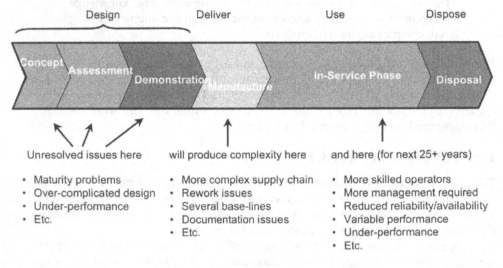

Figure 1: CADMID Lifecycle for a System Capability

The increase in effort, co-ordination and control is significant; if organisations do not rise to meet this challenge, the effects can be long lasting, as illustrated in Figure 1. This incorporates the CADMID lifecycle, well known in the UK defence industry and originally specified by the Ministry of Defence in the UK. The diagram illustrates possible knock on effects of earlier unresolved problems.

2 Decision Making Systems

Decision-making is affected by a number of things. There will be external environmental and commercial pressures, but there are also internal effects and pressures. Many of these, such as established processes, organisational structures and incentivisation policies are typically determined by the company's overall strategy (for example, having an aim to work 'faster, better, cheaper' than competitors). It is important that strategy is developed and implemented in an arena of good governance, otherwise the endeavour to make good decisions will be academic and detrimental emergent behaviour will be a near certainty.

Note the use of the term Decision-Making Systems (DMS): this includes:

- Agents – software or human based, who are involved with decisions,
- Activities – the decision-making activities which enable decisions to be made,
- Infrastructure – which enables decision-making and may include computer based support,
- Knowledge and information – necessary for decision-making.

DMS will also of course be affected by time (time available in which to make the decision, time by when the decision must be made, time when the output or effect of the decision is realised, and the duration of the decision) and the style or process of decision-making.

In the view of the authors, DSS (Decision Support Systems) form part of the overall DMS. In much of the literature [e.g. Silver 1991, Finlay 1989] the term DSS or decision support is considered to extend only to computer based tools. However, this research widens that definition to include any form of support or guidance, which may be computer based, or from a human source, such as in Communities of Practice (CoPs) [Coakes and Clarke 2006, Wenger 2000].

3 The Effects of Complexity

"Complexity is really just reality without the simplifying assumptions that we make in order to understand it." [Allen et al. 2005]

This is a very neat encapsulation of complexity; however, for the purposes of this research, we have adopted a definition very similar to that of Rycroft and Kash [1999]:

> "A system exhibits complexity if it is composed of many integrated entities of heterogeneous parts, which act in a coordinated way and whose behaviour is typically nonlinear."

Issues of complexity and emergence typically are often considered from two pools of interest. Firstly, there are those researchers concerned with the emergence of order from disorder, Conway's 'Game of Life' being a good example of this [Gardner 1970]. A different approach, attuned to an organisational perspective, is to consider the emergence of a different order (or disorder) from an established or planned order. This research is positioned in this second pool of interest, since it is likely to be of most interest to those in the service industries.

Within this latter class of complexity, it is also possible to identify two sources of complexity. Intrinsic complexity is that which arises from what is being attempted in the process or in the problem being addressed. Secondly, there is induced complexity due to the way the organisation is structured and organised to deal with the problem.

3.1 Intrinsic Complexity

Intrinsic complexity arises largely from the interactions between components in a system, both intended and unintended. There are a number of potential solutions/ issues to consider which may help *reduce* intrinsic complexity (it should be realised that *elimination* of intrinsic complexity is impossible):
- Modularity in design can enable containment of complexity,
- System maturity, for example state of knowledge and quality of knowledge management,
- Architecture for core components/ system facets and rigid adherence to standards,
- Clear understanding of the problem context,
- A stable project environment, in terms of budget, timescales, client coherence, partners etc. l(has the biggest impact, but is the most difficult to achieve and unlikely to occur.

3.2 Induced Complexity

Induced complexity is perhaps the most prevalent and powerful reason why development and change projects do not deliver as planned.

Inducing extra complexity into the delivered capability via inappropriate project management (or conversely avoiding or reducing induced complexity via appropriate project management) is discussed widely in literature [de Meyer, Loch and Pich, 2002; Koskela and Howell, 2002; Williams, 2005].

However, as these authors indicate, the style of project management is the result of organisational culture, role assignment and knowledge of the problem area [Sinclair, 2007] but is not the only source of induced complexity; the client, the legal framework, and many other entities can create a project environment that is almost certain to trigger induced complexity.

4 Ongoing studies and Future Work

Initial empirical work included an in-depth literature review and case studies; the first set looked at student engineering groups, investigating the identification of key decision points and the mechanisms of decision making and decision support employed; the second set were based in the aerospace and construction sectors.

Space does not permit a full description of these studies, but will look at one key output– a DMS Framework (see figure 2). The aim of the framework is to help orient organisations with regards to their issues, especially with regards to DMS. The increased information provided will allow the organisations to more appropriately configure their own DMS. It allows data to be captured and represented in a common format and an exemplar of the framework, is shown in figure 2 partially populated with issues, items etc. drawn from accident reports, the literature and the case studies themselves. The row headings are the component parts of a DMS, as explained earlier in the paper. The column headings are aspects of the overall system or system of

	1. Contextual Variables	2. Environmental Variables	3. Organisational Culture	4. Level of DM
A. Agents/ roles	Poor agent allocation or definition for particular part of the lifecycle – e.g. finance representative not present even though budgetary decisions are to be made.	Agents in place cannot deal with external variables, e.g. are unable to act on changing inputs.	Poor agent allocation due to organisational structure.	Agents do not have correct authority to make decisions.
B. Activities	Inappropriate activities for the particular part of the lifecycle.	Correct activities are not in place to deal with external influences, e.g. not configured to take advantage of external opportunities.	Mandated activities prevent effective collaboration.	Inappropriate activities for the level of decision-making.
C. Infrastructure	Inappropriate or non-availability of infrastructure for phase of lifecycle.	Inappropriate infrastructure to deal with external influences.	Inappropriate infrastructure prevents effective collaboration.	Inappropriate infrastructure for the level of decision-making.
D. Knowledge and Information	Inappropriate information and knowledge for phase of lifecycle, e.g. detailed requirements not available during testing.	Inappropriate or unavailable knowledge or information regarding external influences.	Inappropriate information or knowledge preventing effective risk management.	Inappropriate information or knowledge for level of decision making, e.g. detailed technical information at a strategic level.

Figure 2: Example DMS framework with example contents

systems, which could affect or be affected by the DMS. These are:

- Contextual variables - internal issues, for example, what stage are you at in the ELC (Engineering Life Cycle) and what impact does this have?
- Environmental variables- external influences, for example legislation and health and safety.
- Organisational culture, issues such as: Power distance – structure and empowerment; Risk – how much risk are the agents and the organisation willing to take? What impacts are there on the risk management system? Is it possible to take too much or too little risk?; Regimentation – what is mandated? How are things 'usually done'? What is standard practice for the organisation?; Collaboration – individual work vs. collaborative work; Level of decision-making (DM) - Strategic, tactical or operational.

This framework will be matured as the research continues and will be reported in more detail in the future.

5   Conclusion

Across all domains there is an exponential growth in the complexity of a range of long-life systems that comprise our industrialised society. We need a step-change in knowledge of how to design, integrate, operate and evolve systems a) that are not fully understood by all stakeholders b) whose behaviour may not be fully predictable and c) which function in an environment which cannot always be controlled. Complexity is an inherent feature of these systems and is characterized by:

- A (usually large) number of (usually strongly) interacting individual components of the system, and (probably) evolving interaction between the system and its environment, and;

- The requirement for the system to adapt to change in a way that does not have adverse effects on the system's usefulness, nor its ability to operate within a defined envelope of appropriate measurable parameters.

Although provider companies have in-depth expertise and knowledge about characteristics and behaviour of individual system building-block components, there is a lack of understanding of the multiple, non-linear connections and dependencies among components - and of the way that they may self-organise, or co-evolve within a constantly changing environment over its lifecycle. The response of the complex system, therefore, cannot be adequately planned, understood, nor anticipated. Understanding these issues is an incremental process. Developing tools and approaches to aid decision making to ensure the integration of hierarchies of components, sub-systems etc. and to at least allow the properties of a complex System of System (SoS) to be bounded, would be a valuable first step.

The end goal for this research is described as being able to identify key decision points and the appropriate configuration of DMS capable of delivering and servicing complex long-life systems, which must function with a high degree of uncertainty in the system operating space. The DMS must cope with:

- The containment of undesirable emergent behaviour in the delivered system,

- Reduced timescales with controlled costs within the delivery system,

- On-going customer requirement for improved flexibility, interoperability and supportability of the system through its life cycle.

Decision support must be extended beyond rules and guidelines for each and every decision. Decision makers must be enabled with the tools and knowledge to make them aware of the likelihood of emergent behaviour of the systems and help them make resilient decisions. Any tools must be flexible enough to cope with the potential variety of events and robust enough to withstand the challenge of complexity.

Bibliography

[1]   Allen, P.M., Boulton, J., Strathern, M., and Baldwin, J. (2005) The implications of complexity for business process and strategy. In K. Richardson (Ed.), Managing organisational complexity: philosophy, theory and application (pp. 397-418)

[2]   Ashby, W.R. (1956) An introduction to cybernetics, London: Chapman & Hall

[3]   Coakes, E., et al. (2006). Encyclopaedia of communities of practice in information and knowledge management. London: Idea Group.

[4]   de Meyer, A., Loch, C.H., and Pich, M.T. (2002) Managing project uncertainty: from variation to chaos Sloan Management Review 43(2), pp. 60-67

[5]   Finlay, P. (1989). Introducing decision support systems. Oxford Manchester: NCC Blackwell.

[6]   Gardner, M. (1970). The fantastic combinations of John Conway's new solitaire game "life", Scientific American, 233, October, pp. 120-123

[7]   Koskela, L., and Howell, G. (2002) The theory of project management explanation to novel methods Paper presented at the Proceedings of 10th Conference of the International Group for Lean Construction, 6-8 August 2002, Gramado, Brazil

[8]   Rycroft, R.W. and Kash, D.E. (1999), The complexity challenge London: Pinter

[9]   Silver, M. S. (1991). Systems that support decision makers: description and analysis. Chichester: John Wiley & Sons.

[10]   Sinclair, M.A. (2007) Ergonomics Issues in Future Systems. Keynote paper to Ergonomics Society Annual Conference, Nottingham, April 2007. Paper to be published in Ergonomics, 2007

[11]   Wenger, E. (2000). Communities of Practice and Social Learning Systems. Organisation articles, 7(2), pp. 225-246.

[12]   Williams, T.M. (2005) Assessing and moving on from the dominant project management discourse in the light of project overruns IEEE Transactions on Engineering Management 52(4) pp. 497-508

Chapter 14

Rapid Software Evolution

Borislav Iordanov
Kobrix Software Inc.
biordanov@acm.org

1 Introduction

The high complexity of modern software systems, described more than twenty years ago in the well-know paper by Fred Brooks [Brooks 1986] has become proverbial amongst practitioners. While the software engineering community has accepted and learned to cope, albeit in a limited way, with what Brooks termed *essential difficulties*, i.e. intractable obstacles due to the nature of software itself, there is a consensus that the ultra large-scale systems of the future call for a fundamental change in our understanding and practice of software construction [Northrop 2006].

In this paper, we present a new approach to software development, based on the idea of evolutionary engineering [Bar-Yam 2002, 2003] and describe a concrete platform facilitating its implementation. First we outline the essential high-level features of our proposal - a live, multi-paradigm, knowledge-driven, distributed network of operating environments where programmers and users interact within the same environment on a large scale. Next we describe the distributed memory architecture at the foundation of the platform - a generalized hypergraph data structure with a universal representation schema and discuss some of the low and high-level evolutionary dynamics based on it. Finally, we illustrate some social ramifications should such a platform be adopted.

2 Evolving Software

In recent years parallels between the living world and software programs have become recurrent. However, the practice of software development remains the fruit of historical accidents, largely untouched since the early foundational days of the field, when the theoretical focus was chiefly on algorithmic efficiency and the practice of artifact construction inspired by industrial engineering processes that mandate a substantial amount of forethought due to the prohibitive costs of design flaws. Attacking the complexity of large projects through divide-and-conquer methods such as abstraction and modularity is the *modus operandi* of the engineering community. Even when inspired from living systems, researchers usually focus on defining the proper abstractions and modularization boundaries [Gabriel 2006, Imbusch 2005, Fleissner 2006]. This is only natural since engineering tasks are tackled top-down, starting from the problem and sub-dividing it into simpler problems. Unfortunately, the strategy does not scale as argued extensively in [Bar-Yam 2002]. Software systems suffer from the rigidity of their abstractions and from the unpredictable, non-linear interactions between their modularized components. What can be done?

When faced with the task of creating highly complex systems where top-down decomposition is not possible, one realizes a basic contradiction: emergent behavior is by definition unpredictable and therefore contrary to the engineering mandate of building systems with well-defined functions.

How does one achieve goal-directed emergence? The answer is through biological evolution. Classical engineering achieved its success by turning a descriptive theory about the world (physics) into a prescriptive tool. Similarly, we expect the theory of evolution to yield successful emergent systems when applied in a prescriptive engineering context.

Goal-directed evolution presupposes a fitness function, a way to measure the relative merits of organisms. It has been successfully applied at a low-resolution scale, where the fitness function is some easily computable numerical quantity, in the case of genetic algorithms. However, evolving large end-user programs from bit strings would be clearly impractical. At the opposite spectrum we have a market-based ecosystem of software programs, each grown by following a classical engineering lifecycle. There the fitness function is defined by collective human feedback and the process works, but with significant shortcomings: first, there must be an actual market for a particular application domain (frequently not the case for highly specialized, commissioned projects); second, the evolutionary time-scale is large (actually a function of the complexity of the software); third, the granularity is too coarse (we frequently witness the disappearance of high-quality features when the whole program loses the market battle).

A more promising approach to software evolution would frame it as a middle ground, between the above two extremes, and within software's natural habitat: the cognitive landscape of human knowledge, interaction and information exchange. In other words, the right scale of evolutionary units is the one at which human cognition operates, the level of abstraction permitted by current technology and where entities can be recombined in a meaningful way. This could mean a single programming unit such as a data type or a procedure, or a piece of data upon which behavior is based in a clearly recognizable way. It is primarily humans who make up the environment where computer programs evolve and therefore all forms of human participation, from the hard-core programmer to the neophyte end-user, should be woven together into the same computing medium.

The sharing of the same interactive medium between consumers and producers of software is a key point if feedback must operate at fine granularity. Programs are usually conceived as standalone executable files compiled out of source code representation in some high-level language. The abstract models present in the source code are lost during the compilation process and much engineering effort is spent in categorizing software behavioral decisions into *design time*, *compile time*, *deployment time, and run time*. However, alternative models where the compile time vs. run-time distinction is blurred have emerged over the years, the most notable perhaps being one of the first object-oriented languages Smalltalk [Goldberg 1993] and its modern reincarnation Squeak [Squeak]. Another prominent example is the Self environment [Ungar 1987]. Those systems are termed "live" systems because they are entirely made up of persistent live objects, completely exposed and modifiable by the user in the very form in which they were originally created. Our platform falls in the category of live systems, but with several essential differences.

First, we avoid tying the programming task to any meta-model, programming paradigm or language. Rationale: programmers with diverse backgrounds should be able to participate; software problems yield themselves better to one paradigm or another; diversity is good.

Second, in order for fine grained software artifacts to replicate, diversify, be selected for or against, the platform is distributed in nature, yielding a decentralized network of such live systems.

Third, knowledge representation capabilities are natively incorporated for the following main reason. In classical monolithic systems, the gluing of components is to a large extent based on the programmer's hidden intent behind each part (including, but not limited to the so called "implementation details" that software designers try to make irrelevant). In an evolutionary system where much variation and recombination of sub-components is common and where, in addition, such variation is to be primarily induced by contextual particularities of the environment, semantic metadata about both artifacts and environment is of essence for management features such as compatibility/applicability checks, exception handling and the like.

Finally, we note that human cognition is capable of operating at different resolution scales, depending on context. Therefore, the units of evolution should span different organization levels as well. A crucial component of the platform, allowing it to meet all of the above requirements, is a highly flexible, structured, distributed memory model which we now describe.

3 HyperGraph Structured Memory Domain

The memory model is based on the most fundamental principle of organization - aggregating two or more entities. In formal terms, it is a generalized hypergraph. A hypergraph is a graph where edges may point to more than two nodes. The generalization further allows edges to point to other edges. Edges and nodes are thus unified into the single notion of a hypergraph atom where each atom has an arity - the number of atoms it points to - which is a number ≥ 0. An atom with arity 0 is called a *node* while an atom with arity > 0 is called a *link*. The atoms a link points to are called its *target set*. This structure was invented and proposed as a cognitive model for artificial general intelligence by Ben Goerztel [Goerztel 2006]. In addition, atoms are typed and carry a value as a payload. However, the connection of an atom with the rest of hypergraph is independent of its type and value. The type system itself is embedded in the hypergraph structure and it is completely open and able to accommodate virtually all computer languages. Each type is also a hypergraph atom that is typed in turn. Types of types are called *type constructors*. The system is bootstrapped with a set predefined types and type constructors from which other types are built and evolved. The whole storage model is open and it is based on the following organizational schema:

Atom	→	Type Value TargetSet
TargetSet	→	Target1 Target2 ... TargetN
Type	→	Atom
Value	→	Part1 Part2 ...
Value	→	RawData

Each element in the above grammar except RawData, which is simply a sequence of bytes, is a UUID (Universally Unique Identifier). This ensures a globally unique identity of all hypergraph elements and provides an universal addressing schema, thereby allowing the memory to be distributed into a decentralized network of local environments.

By convention, atoms are mutable while their values are not. This means that an atom can have its value replaced, or its target set modified while still preserving its identity and therefore the network structure of the hypergraph. There's no preset level of granularity for hypergraph atoms. Uniformity of the data representation schema has proven of essence in many systems, and we stick to that principle. Atoms range from simply-valued (e.g. strings, numbers) to complex records, self-contained user interface components, logical terms (e.g. represented as links between sub-terms), any kind of executable (in source or compiled form) or documentation resource[1]. Therefore, all those familiar software artifacts share the same representation medium and can therefore freely refer to each other opening the door for arbitrary layers and levels of organization. For instance, semantic information about executable entities is readily available to a run-time environment and interpreted to enforce constraints and manage dependencies.

In the following section, we illustrate how the hypergraph memory domain foundation is used to implement an evolutionary engineering software platform.

4 Dynamics of Evolutionary Software Construction

4.1 Overview

The platform is based on the idea of a large number of interconnected operating environments, which we call *niches*, linking together end users and developers within a single computational medium. A niche is bound to a single hypergraph instance. The environment is bootstrapped as a standard application based on the current user's

---

[1] It is not hard to imagine how all those examples can be represented with the general representation schema above. We omit such details for brevity.

configuration (essentially from where the user left of during the last run). Each instance is connected to other instances so that hypergraph atoms can be shared, replicated or distributed in a peer-to-peer fashion. Clusters of hypergraphs may exist within a single organization, or span geographic locations and serve as a collaborative medium for teams and programmers. We expect the topology of the niche network to resemble a scaled-down version of the topology of the internet.

The units of evolution are atoms. It is atoms that get replicated, vary and are selected for or against. Because an atom is a handle to an entity at any level of organization, the granularity of the elements of evolution is not mandated by the platform. Instead, it is contextual for a specific application, development task and/or organization. Selective pressures operate in two ways:

1. On atoms, where participants decide what atoms to keep in their niche. For instance, programmers decide what software components they need while end-users decide on "end-user" software artifacts, or the system itself decides to eliminate atoms that are no longer in use.

2. On an atoms' informational import (i.e. its type, value and target set). Recall that atoms always keep their identity while their informational import can vary between niches. This mechanism is important for establishing a boundary between a reference (the atom UUID) and its exact interpretation which is allowed to vary while still preserving referential integrity.

A user may be connected to such a cluster solely for consumption purposes without ever participating in variation of the population of atoms as created by programmers. On the other hand, more sophisticated users may induce variation by modifying data that drives behavior, for example by changing a significant "configuration setting" or encoding expert knowledge in some declarative logical form or creating a macro command.

4.2 Low-Level Dynamics

We now describe a few low-level scenarios of atom replication and variation. Assume a peer-to-peer communication layer between two given niches N and M and suppose that N promotes/publishes/offers an atom to M. This publishing can be triggered manually or automatically through some preset negotiation protocol, or simply be part of a larger update process. The atom may encompass functionality at any level - from the background color of a window to a word processing program. If the atom doesn't exist at M, it is simply transmitted and it's up to N to keep it or remove it at a later time. If the atom exists at M and it has the same informational import, nothing is done. If, on the other hand, the atom exists at M and has a different informational import, it is updated recursively by updating its type and target set. The old version is preserved under different UUID and tagged (via a hypergraph link) as being overwritten by this particular update. Thus an update can be completely reversed in the future - selected against.

What kind of variation of a given atom may we expected? The simplest case is when its value has changed. A more elaborate one is a type change since it may result in a different run-time instance of the atom. In this case approaches such as [Evans 1999] can be adopted. But atoms in the target set can change as well. For example, one of the mechanism responsible for the efficiency of the evolutionary process in the living world is sexual reproduction. This gives rise to generational variation through recombination (crossover) of existing characteristics. Hypergraph atoms exhibit such phenomena by mutating their target sets. For instance, suppose an atom A has type T, value V and a target set consisting of atoms X, Y and Z:

$$A \rightarrow T, V, X, Y, Z$$

For concreteness, one may imagine that A is a software module and X, Y and Z are sub-modules. Suppose further that at a certain point in time A is identical across a topology of three niches N_1, N_2 and N_3 where N_1 is connected to both N_2 and N_3, but N_2 and N_3 are not connected. Now, eventually Y turns into Y' at N_2 and Z turns into Z' at N_3. The niche N_1 can then acquire both changes sequentially thus recombining the two versions - a crossover effect. It may turn out that when both changes are combined, this has an adverse overall effect on the function of A in which case one, one or both updates can be rolled back. If, on the other hand, all goes well, N_1 may redistribute its aggregated version back to N_2 and N_3 which may in turn reverse the change because it fails within their context, or propagate it further.

The nature of the representation in the above example was essential for the possibility of crossover variation. For this reason, programmers are encouraged, but not forced, to create representations that rely on the hypergraph structure. Such representations may evolve, of course.

Note that those scenarios don't assume human communication. When a target niche is updating an existing functionality and it has a battery of automated tests to ensure quality and/or measure improvement, no human involvement is necessary.

4.3 High-Level Dynamics

At a larger scale, we note two important aspects of evolutionary dynamics. The first is stability vs. chaos. Given highly structured representations where dependencies are explicit in the graph structure, the system can be very helpful in tracking and resolving them. However, unpredictable inconsistencies can occur due to atom inter-dependencies that are not explicitly represented in the graph. And here lies one of the main difficulties: many failures in software are latent, subtle and the result of side-effects. This leads to (what is perceived as) chaotic behavior. The proposed platform does not attempt to solve the complexity problem by encoding every possible dependency between entities in a graph structure. Rather, it makes the process of parallel exploration of possibilities more efficient, in part by allowing for failure, experimentation and improvement at a fine-granular level. In this context, various niches assume roles similar to current development/deployment processes. For instance, in a mission/business critical environment, a "staging" niche would adopt a change before it is broadcast to "production" niches while a cluster of development niches will be highly fluctuating. A complementary aspect to niche role partitioning are atom update negotiation protocols which range from automatic synchronization to controlled, manually triggered requests. Stability then emerges at the local level out of social pressures (usage, tolerance) as well as computational resources constraints.

The second aspect is uniformization vs. diversification. Uniformization is the result of such pressures as predictability - one wants to know what to expect from a piece of software, common understanding - one wants to be able to share information such as training materials and costs reduction - cheaper to reuse something existing than building anew. Diversification is the result of adaptation to a context of use where context includes thing like business processes, individual preference, computational resources available. Uniformization and diversification occur at all levels of organization. For example, there's little reason for different versions of a sorting routine while there are good reasons for different implementations of an abstract data type (such as a key-value lookup structure) that perform better depending on the nature of the data. Similarly, there's little reason for the implementation of a pull-down menu component to vary from system to system, but there is good reason to vary the implementation of a command navigation system in the context of users with disabilities.

5 Social Implications

The long-term vision behind our effort is a large network of niches connecting participants with any conceivable software needs and technical knowledge. A sweeping fine-grained decentralization of the production of software artifacts would certainly alter its economic dynamics. The current open-source model of a large body of freely available software with a service economy around it may well serve as a precursor, albeit a mediocre one. Open-source software is still centralized and it follows standard engineering practices. In addition, code ownership remains the *de facto* reality. By contrast, our proposal entails the eventual abundance of interconnected fine-grained information artifacts with collective ownership.

The live aspect of the platform encourages more active participation from traditional end-users. Enforced or emergent variability within the software artifact space breaks the traditional information sharing about and trust in the predictability of immutable monolithic programs with identical behavior everywhere. One consequence could be the birth of localized services or "artisans" offering technical expertise in niche contextualization and absorbing the adverse effects of failed evolutionary experiments.

As a more down to earth example, consider the widespread software business model where a company licenses a product and offers customization services. The internal organization of such companies usually follows a pattern where the product is developed by a core team of highly skilled engineers with satellite teams providing solutions in the form of a set of services and customizations to different clients. This organization can be depicted in Figure 1. The product team interacts with several solutions teams and each solution team interacts with several customers in what amounts to a strict hierarchical organization. All customers share the same core product capabilities. Customizations are possible only when variability has been explicitly introduced upfront. When this is not the case, customer demands drive a core future to be detached as customizable. In fact, there's pressure for the core product to incorporate enough variability so as to meet everybody's needs, thereby augmenting its complexity and creating the bottleneck so characteristic of other hierarchical structures. By contrast, in an evolutionary platform there is no core product at all (see Figure 2). The company instead maintains a cluster of niches, each loosely adapted to particular customers. Solution teams form a decentralized collaboration network where each manages a niche tailored towards specific business needs. The variety of the company's business environment is much better represented in the new arrangement as suggested by the Law of Requisite Variety [Ashby 1956].

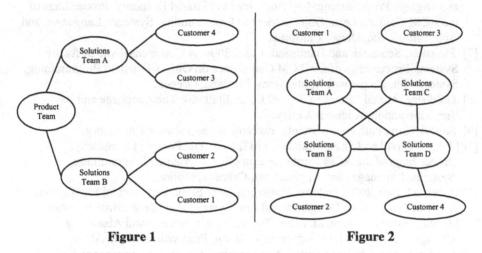

Figure 1   Figure 2

6 Final Remarks

In this work, our main assumption has been that the key to mastering the complexity of the software systems of the future is not the "right" set of abstractions, for it is their endless deluge that generates the complexity in the first place. Rather, it is in the process of goal-directed emergence as enacted through evolutionary engineering. Our ongoing work on the hypergraph memory domain and run-time environment for the platform (both free and open-source) can be found at http://www.kobrix.com.

7 Bibliography

[1] Brooks, Frederick, 1986 No Silver Bullet: Essence and Accidents of Software Engineering. s.l. : Information Processing.
[2] Northrop, Linda et al., 2006 Ultra-Large-Scale Systems. s.l. : Software Engineering Institute, Carnegie-Mellon.
[3] Bar-Yam, Yaneer, 2002, Large Scale Engineering and Evolutionary Change: Useful Concepts for Implementation of FORCEnet. s.l. : Report to Chief of Naval Operations Strategic Studies Group, available at

http://necsi.org/projects/yaneer/SSG NECSI 2 E3 2.pdf.

[4] Bar-Yam, Yaneer, 2003, When Systems Engineering Fails — Toward Complex Systems Engineering.. Piscataway, NJ : IEEE Press, International Conference on Systems, Man & Cybernetics. Vol. 2, pp. 2021-2028.

[5] Gabriel, Richard P and Goldman, Ron, 2006, Conscientious Software, Proceedings of the ACM Conference on Object-Oriented Programming, Systems, Languages and Applications, Portland, Oregon.

[6] Imbusch, Oliver, Langhammer, Frank and von Walter, Guido, 2005, Ercatons and Organic Programming: Say Good-Bye to Planned Economy, Proceedings of the ACM Conference on Object-Oriented Programming, Systems, Languages and Applications, San Diego, California.

[7] Fleissner, Sebastien and Baniassad, Elisa, 2006, A Commensalistic Software System, Proceedings of the ACM Conference on Object-Oriented Programming, Systems, Languages and Applications, Portland, Oregon.

[8] Goldberg, A. and Robinson, D, 1993, Smalltalk-80: The Language and Its Implementation, Addison-Wesley.

[9] Squeak (Smalltalk environment), available att http://www.squeak.org.

[10] Ungar, David and Smith, Randall, 1987, Self: The Power of Simplicity, Proceedings of the ACM Conference on Object-Oriented Programming, Systems, Languages and Applications, Orlando, Florida.

[11] Goertzel, Ben, 2006, Patterns, Hypergraphs & Embodied General Intelligence, IEEE World Congress on Computational Intelligence, Vancouver, British Columbia

[12] Evans, Huw and Dickman, Peter. Zones, 1999, Contracts and Absorbing Change: An Approach to Software Evolution, Proceedings of the ACM Conference on Object-Oriented Programming, Systems, Languages and Applications, Denver, Colorado

[13] Ashby, Ross W., 1956, An Introduction to Cybernetics, Chapman & Hall, London.

Chapter 15

Structural Changes and Adaptation of Algal Population under Different Regimens of Toxic Exposure

Valentina Ipatova, Valeria Prokhotskaya, Aida Dmitrieva
Department of Biology
Moscow State University MV Lomonosov
viipatova@hotmail.com, plera@mail.ru

1. Introduction

Algae are the principal primary producers of aquatic ecosystems. Modern chemical residues from water pollution (such as pesticides in surface and ground waters, antibiotics, chemical substances of military use, heavy metals, oil, oil products, etc.) are a challenge to survival of microagal populations. Growth of many species was restricted even by micromolar concentrations of such xenobiotics.

Laboratory populations of microalgae are widely used as sensitive test object for the evaluation of the phytotoxicity of chemicals and wastewater streams. Cell populations of microalgae are complex systems with resistant and sensitive cells. When pollutants are added to a dense microalgal culture, the cell density will be reduced after a few days due to the death of sensitive cells. However, after further incubations, the culture will sometimes increase in density again due to the growth of cell variant, which is resistant to the contaminants. Numerous studies have shown that heavy metals are extremely toxic to microalgae in both laboratory cultures and natural populations. It has also been reported that microalgae from contaminated sites appear to have adapted to high metal concentrations whereas algae from unpolluted sites remain sensitive [Knauer 1999].

Rapid adaptation of microalgae to environmental changes resulting from water pollution has been demonstrated recently [Costas 2001, López-Rodas 2001]. Unfortunately, the evolution of microalgae subsequent to a catastrophic environmental change is insufficiently understood. Little is known about the mechanisms allowing algal adaptation to such extreme conditions. Within limits, organisms may survive in chemically-stressed environments as a result of two different processes: physiological adaptation (acclimation), usually resulting from modifications of gene expression; and, adaptation by natural selection if mutations provide the appropriate genetic variability [Belfiore 2001]. Because physiological adaptation is bounded by the types of conditions commonly encountered by organisms, it remains for genetic adaptation to overcome extreme environmental conditions [Hoffmann 1991].

The changes of population structure of freshwater green alga *Scenedesmus quadricauda* and marine diatom alga *Thalassiosira weissflogii* were studied under different regimens of heavy metal (chromium) exposure. Adaptation of the algae to growth and survival in an extreme environment was analysed by using an experimental model. The main aims of this work were: (1) to estimate the effect of chromium contamination on microalgal populations under different regimens of chromium addition; (2) to determine the nature and origin of chromium-resistant cells that arise; (3) to estimate the mutation rate from chromium sensitivity to chromium resistance.

2. Materials and Methods

The culture of green chlorococcal alga *Scenedesmus qudricauda* (Turp.) Breb. (strain S-3) was grown non-axenically in Uspenskii medium N1 (composition, g/l: 0.025 KNO_3, 0.025 $MgSO_4$, 0.1 KH_2PO_4, 0.025 $Ca(NO_3)_2$, 0.0345 K_2CO_3, 0.002 $Fe_2(SO_4)_3$; pH 7.0-7.3) in conical flasks in luminostat under periodic illumination (12:12 h). The culture of diatom alga *Thalassiosira weissflogii* (Grunow) Fryxell et Hastle was grown non-axenically in Goldberg-Kabanova medium (composition, g/l: 0.2024 KNO_3, 0.007105 Na_2HPO_4; mg/l: 0.1979 $MnCl_2$, 0.2379 $CoCl_2$, 0.2703 $FeCl_3$).

Toxicity test: effect of chromium on population growth. We investigated the toxic action of potassium dichromate ($K_2Cr_2O_7$, PD), well known as standart toxicant [Wang 1997], in view of maintenance of a constant dose of chromium per one cell during experiments in order to pass from concentration dependence to dose dependence. The laboratory algal cultures were exposed to increasing concentrations of the toxicant in the long-term experiments in three replicates. The experiments were performed both with single chromium addition at the start of experiment and with multiple additions during exposure time. The periods between toxicant additions approximately corresponded to doubling time for algae so that the dose of the toxicant per one cell was particularly the same as that at the initial day of experiment. The effect of chromium on *S. qudricauda* and *T. weissflogii* was estimated by calculating total cell number, a share of alive, dead and dying cells during exposure time (28 and 21 days, respectively). Cells were counted with a Goryaev's hemocytometer and Nazhotta cytometer under a light microscope. Number of alive, dead and dying dead cells was counted with luminescent microscope Axioskop 2FS (Carl Zeiss, Germany).

For experiment with *S. quadricauda* we used concentration of chromium: 0.001; 0.01; 0.1; 1; 5 and 10 mg/L. Concentration of toxicant in a stock solution was 1 mg/mL

150

(counting per chromium). Initial number of cells after inoculation was 50 000 cells/mL. After that cells grew within 5 days on reaching of a logarithmic growth phase by culture. Number of cells at this moment was $28\text{-}30\cdot10^4$ cells/mL. Experiment was performed in conic flasks in volume of 100 mL, volume of culture in which was 50 mL. We added toxicant to cultures at 0 day of experiment (single addition) and further at 3, 6, 10 and 17 day until necessary concentrations (multiple additions). Frequency of toxicant addition was defined by growth rate of cultures and rate of cell division.

For experiment with *T. weissflogii* we used concentration of chromium: 0.001; 0.01; 0.1 and 1 mg/L. The initial number of cells taken for experimemt was 5 000 cells/mL. Experiment was performed in small phials with 10 mL of culture. Chromium was introduced into growth medium at 0 day of experiment until necessary concentrations (single addition). Further, in one series of culture we did not add the toxicant (conditionally named by us as "control") and in another series chromium was added at 3, 6, 10 and 13 day (multiple additions).

Average growth rate of both cultures (without chromium) was 0.33 division/day. Toxicant was introduced into the growth mediums proportionally to an increase of cell number of *S. quadricauda* and *T. weissflogii* so that the toxicant quantity per one cell (dose) was kept constant.

Fluctuation test: analysis of transformation from chromium sensitivity to chromium resistance. A modified Luria–Delbrück fluctuation analysis was performed as previously described [López-Rodas 2001] in liquid medium to distinguish resistant cells that had originated as a result of random spontaneous pre-selective mutations (prior to chromium exposure) from those arising through acquired post-selective adaptation (during the exposure to chromium).

Two different sets of experimental cultures were prepared with both species of algae. The first set of experiments was performed in 52 (*S. quadricauda*) and 49 (*T. weissflogii*) parallel culture flasks with cell number $N_0 = 200$ cells and $N_t = 2.8\cdot10^4$ (*S. quadricauda*), $N_t = 10^5$ (*T. weissflogii*) cells; and treated with 2.5 (*S. quadricauda*), 1.5 (*T. weissflogii*) mg/L chromium after reaching N_t. For the second set of experiments, 30 aliquots of 10^4 (*S. quadricauda*) and 10^5 (*T. weissflogii*) cells from the same parental populations were separately transferred to flasks containing fresh liquid medium with 2.5 (*S. quadricauda*) and 1.5 (*T. weissflogii*) mg/L chromium. Cultures were observed for approximately 14 days, and the resistant cells in each culture (both in set 1 and set 2) were counted. The cell count was performed by at least two independent observers.

If resistant cells arise by rare spontaneous mutations, each parallel culture in set 1 would have a given probability of generating resistant variants with each cell division. Then, inter-flask variation would not be consistent with the Poisson model. The number of cells from each flask in set 2 would show variation due only to random sampling; variation from flask to flask would be consistent with the Poisson model. If there is rare spontaneous mutation, the variance/mean ratio$_{set1}$ is usually many times higher than the variance/mean ratio$_{set2}$. The method allows estimation of the rate of spontaneous mutation in algae and the rate of appearance of resistant cells. The proportion of set 1 cultures showing no mutant cells after chromium exposure (P_0 estimator) was the parameter used to calculate the mutation rate (μ). The P_0 estimator [Luria 1943] is defined as follows: $P_0 = e^{-\mu\,(N_t - N_0)}$, where P_0 is the proportion of cultures showing no resistant cells. Therefore, μ was calculated as: $\mu = -\text{Log}_e P_0 / (N_t - N_0)$.

3. Results and Discussion

We tried to develope an experimental model of toxic effect using constant toxicant dose per cell during the experiments.

The presented data show (Fig. 1), that at presence of high chromium concentration (1 mg/L and more) the total cell of both species slightly varied or decreased, since the moment of the first chromium addition and down to the end of experiment in comparison with the initial cell number and drastically decreased in comparison wich control without chromium. At toxic influence of such intensity, the dose of chromium per one cell remains practically constant during all term of experiment. Therefore with reference to high concentration of substances it is possible to speak about concurrence of concepts "concentration" and "dose" even if we add the toxicant one time at the beginning of the experiment.

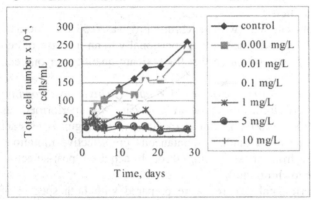

Figure 1. Changes of the total cell number of *S. quadricauda* under chromium exposure (multiple chromium additions).

At medium chromium concentration of 0.1 mg/L number of cells increased, but growth rate of culture has been slowed down in comparison with control one. At low chromium concentration of 0.001 and 0.01 mg/L growth rate of *S. quadricauda* corresponded to the control parameters down to 10 day of experiment, then growth rate have decreased, however by the end of experiment number of cells at presence of these concentrations of chromium has appeared close to the control. Thus, the most sensitive stage at repeated additions of chromium in medium is, apparently, second half of logarithmic growth phase (10-14 day of experiment). As concentration of chromium of 0.001 and 0.01 mg/L are low enough, it is not likely, that they provoke selection of resistant cells. In this case chromium could cause "synchronization" (full or partial) of cultures seaweed by delay or arrest of cellular division at 7-10 day of experiment. After that there was an acclimation of algal cells, and cellular division also was synchronously restored. Thus cultures have reached "control" levels of number of cells.

Thus, at low chromium concentrations of 0.001 and 0.01 mg/L during the experiments with the periodical additions growth rate of *S. quadricauda* was close to the control (without chromium) although the total final concentrations were 3.3-3.4 times more than initial ones.

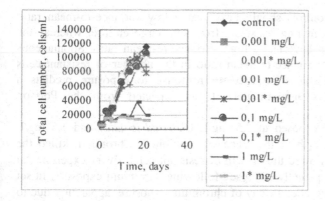

Figure 2. Changes of the total cell number of *T. weissflogii* under chromium exposure: single addition at the start of experiment and multiple (*) additions at 0, 3, 6, 10 and 13 day of experiment.

The final cell number of *T. weissflogii* was slightly decreased in the presence of 0.001 mg/L chromium and was reliably smaller in the presence of 0.01; 0.1 and 1 mg/L chromium during the multiple intoxication as compared with the single one (Fig. 2).

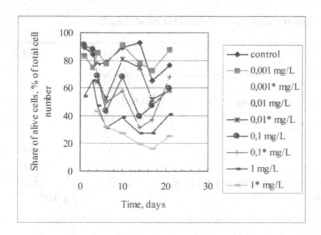

Figure 3. Changes of share of alive cell of *T. weissflogii* under chromium exposure: single addition at the start of experiment and multiple (*) additions at 0, 3, 6, 10 and 13 day of experiment.

The share of dead and dying cells was slightly higher at the multiple intoxication than at the single one (Fig. 3) during experiments with both species (data for *S. quadricauda* are not presented).

We have determined earlier [Prokhotskaya 2006] the number of resistant cells within the heterogeneous *S. quadricauda* population under triple chromium 3.5 mg/L intoxication during 90 days. In spite of the long-term exposition with the toxicant some algal cells remained alive. Their number was 5-6 % of initial population density.

In the present study we have analysed the spontaneous occurrence of chromium-resistant cells in cultures of chromium-sensitive (wild-tipe) cells of *S. quadricauda* and *T. weissflogii*. Modified Luria-Delbrück fluctuation analysis with algae as experimental organisms [Luria 1943; López-Rodas 2001] was used to distinguish between resistant cells arising by rare spontaneous pre-adaptive mutations occurring randomly during replication of organisms prior to the incorporation of chromium and chromium resistant cells arising through post-selective adaptation in response to chromium and, subsequently, to estimate the rate of occurrence of resistant cells.

On the base of hypothesis that adaptation to chromium occurs by selection on spontaneous mutations, the controls should have had a low variance-to-mean ratio consistent with the error in sampling resistants from one large culture, whereas the fluctuation test cultures should have had a high variance-to-mean ratio. Spontaneous mutation thus predicts a high variance-to-mean ratio in the number of resistant cells among cultures, whereas resistance acquired in response to exposure predicts a variance-to-mean ratio that is approximately 1, as expected from the Poisson distribution.

When algal cultures were exposed to 2.5 mg/L (*S. quadricauda*) and 1.5 mg/L chromium (*T. weissflogii*), growth of the algae was inhibited. Chromium killed the wild-type sensitive cells but allowed the growth of resistant cells. Every experimental culture of both sets 1 and 2 apparently collapsed following chromium exposure. In set 1, only some cultures recovered after 14 day of chromium exposure, apparently due to the growth of chromium resistant cells (recovered cultures increased their cell number compared to the control level). A high fluctuation in set 1 (in contrast with the scant variation in set 2) was found in both species (Table 1, 2), which indicated that the high variance found in set 1 cultures should be due to processes other than sampling error.

Table 1. Fluctuation analysis of resistant variants in *Scenedesmus quadricauda*

	Set 1	Set 2
No. of replicate cultures	52	30
No. of cultures containing the following no. of resistant cells/mL:		
0	45	0
$0-2 \times 10^4$	2	0
$2 \times 10^4 - 10^5$	5	30
$>10^5$	0	0
Variance/mean (of the no. of resistant cells per replicate)	61.5	3.2
μ (mutants per cell division)	5.2×10^{-6}	

Table 2. Fluctuation analysis of resistant variants in *Thalassiosira weissflogii*

	Set 1	Set 2
No. of replicate cultures	49	30
No. of cultures containing the following no. of resistant cells/mL:		
0	36	0
1-1300	4	0
1300-5000	9	30
>5000	0	0
Variance/mean (of the no. of resistant cells per replicate)	16.8	0.95
μ (mutants per cell division)	3.1×10^{-6}	

The data from a fluctuation test were used to calculate a spontaneous mutation rate per cell division using the proportion of cell cultures that exhibit no mutants at all [Luria 1943]. The estimated mutation rates (μ) using the P_0 estimator were $5.2 \cdot 10^{-6}$ and

3.1·10^{-6} mutants per cell division in *S. quadricauda* and *T. weissflogii*, respectively.

The data of this study correspond to the results of other work carried out on understanding algal adaptation to anthropogenic chemical water pollutants [Costas 2001; López-Rodas 2001; Baos 2002; García-Villada 2002; Flores-Moya 2005]. The mutation rate from 3.1·10^{-6} to 5.2·10^{-6} mutants per cell per generation was the same order (or one order lower and higher) of magnitude found for the resistance to several pollutants in other cyanobacterial and microalgal species. The presence of resistant cells in the populations of algae is regulated by the recurrent appearance of mutants and their elimination by selection, yielding an equilibrium frequency of 3-5 resistant cells per 10^{6} cell divisions. This fraction of resistant mutants is presumably enough to assure the adaptation of algal populations to catastrophic water contamination, since the algal natural populations are composed of countless cells. Nevertheless, mutations usually imply an energetic cost that may affect the survival of adapting populations [Coustau 2000], as it has been demonstrated by a decreased growth rate in resistant cells compared to growth rate in sensitive ones [Flores-Moya 2005; López-Rodas 2007]. Thus, resistant cells could develop in freshwater ecosystems polluted with the toxicants, but their contribution to primary production will be significantly lower than that occurring in pristine ecosystems with sensitive cells.

4. Conclusion

The present study is a simple model of algal adaptation to stressful environments. Our results suggest that rare preselective mutants can be sufficient to ensure the adaptation of eukaryotic algae to extreme natural habitats. These values are low (~10^{-6} mutants per cell division). Such mutation rate coupled with rapid growth rates, are presumably enough to ensure the adaptation of microalgae to water contamination. The resistant cells arise randomly by rare spontaneous mutation during replication of cells prior to the addition of the contaminant. Resistant mutants are maintained in the absence of contaminants as the result of balance between new resistant cells arising from spontaneous mutation and resistant cells eliminated by natural selection, so that about 3-5 chromium-resistant mutants per million cells are present in the absence of chromium. Within limits microalgal species should survive in polluted environments as a result of physiological adaptation. With increasing concentrations of contaminants, however, physiological adaptation is not enough, but the genetic variability of natural populations could assure the survival of at least some genotypes [Mettler 1988]. Genetic variability in natural populations is the most important guarantee of surviving most environmental changes [Lewontin 1974; Mettler 1988]. Some populations are being exposed to new xenobiotics for the first time. Sudden toxic spills of residual materials can be lethal to microalgae. Rare spontaneous pre-adaptive mutation is enough to ensure the survival of microalgal populations in contaminated environments when the population size is large enough. Adaptation of algal populations to modern pollution-derived environmental hazards seems to be the result of a rare instantaneous events and the result of resistant cells selection within heterogeneous population.

References

Baos, R., Garcíía-Villada, L., Agrelo, M., López-Rodas, V., Hiraldo, F., & Costas, E., 2002, Short-term adaptation of microalgae in highly stressful environments: an experimental model analysing the resistance of *Scenedesmus intermedius* (Chlorophyceae) to the heavy metals mixture from the Aznalcóllar mine spill, *Eur. J. Phycol.*, **37**, 593.

Belfiore, N. M., & Anderson, S. L., 2001, Effects of contaminants on genetic patterns in aquatic organisms: a review, *Mutat. Res.*, **489**, 97.

Costas, E., Carrillo, E., Ferrero, L. M., Agrelo, M., García-Villada, L., Juste, J., & López-Rodas, V., 2001, Mutation of algae from sensitivity to resistance against environmental selective agents: the ecological genetics of *Dictyosphaerium chlorelloides* (Chlorophyceae) under lethal doses of 3-(3,4-dichlorophenyl)-1,1-dimethylurea herbicide, *Phycologia*, **40**, 391.

Coustau, C., Chevillon, C., & Ffrench-Constant, R., 2000, Resistance to xenobiotics and parasites: can we count the cost?, *Trends Ecol. Evol.*, **15**, 378.

Flores-Moya, A., Costas, E., Bañares-España, E., García-Villada, L., Altamirano, M., & López-Rodas, V., 2005, Adaptation of *Spirogyra insignis* (Chlorophyta) to an extreme natural environment (sulphureous waters) through preselective mutations, *New Phytol.*, **165**, 655.

García-Villada, L., López-Rodas, V., Bañares-España, E., Flores-Moya, A., Agrelo, M., Martín-Otero, L., & Costas, E., 2002, Evolution of microalgae in highly stressing environments: an experimental model analyzing the rapid adaptation of *Dictyosphaerium chlorelloides* (Chlorophyceae) from sensitivity to resistance against 2,4,6-trinitrotoluene by rare preselective mutations, *J. Phycol*, **38**, 1074.

Hoffmann, A. A., & Parsons, P. A., 1991, *Evolutionary Genetics and Environmental Stress*, Oxford University Press Inc. (New York).

Knauer, K., Behra, R., & Hemond, H., 1999, Toxicity of inorganic and methylated arsenic to algal communities from lakes along an arsenic contamination gradient, *Aquat. Toxicol.*, **46**, 221.

Lewontin, R. C., 1974, *The genetic basis of evolutionary change*, Columbia University Press, (New York).

López-Rodas, V., Agrelo, M., Carrillo, E., Ferrero, L. M, Larrauri, A., Martín-Otero, L., & Costas, E., 2001, Resistance of microalgae to modern water contaminants as the result of rare spontaneous mutations, *Eur. J. Phycol.*, **36**, 179.

López-Rodas, V., Flores-Moya, A., Maneiro, E., Perdigones, N., Marva, F., García, M. E., & Costas, E., 2007, Resistance to glyphosate in the cyanobacterium *Microcystis aeruginosa* as result of pre-selective mutations, Evol. Ecol., **21**, 535.

Luria, S. E., & Delbrück, M., 1943, Mutations of bacteria from virus sensitivity to virus resistance, *Genetics*, **28**, 491.

Mettler, L. E., Gregg, T., & Schaffer, H. E., 1988, *Population Genetics and Evolution*, 2nd edn. Prentice-Hall, Englewood Cliffs (New Jork).

Prokhotskaya, V. Yu., Ipatova V. I., & Dmitrieva, A. G., 2006, Intrapopulation Changes of Algae under Toxic Exposure, *Proc. Int. Conf. on Complex Systems 2006*, http://necsi.org/events/iccs6/viewpaper.php?id=50.

Wang, W., 1997, Chromate ion as a reference toxicant for aquatic phytotoxicity tests, *Environ. Toxicol. Chem.*, **6**, 953.

Chapter 16

Modularity and Self-Organized Functional Architectures in the Brain

Laxmi Iyer and Ali A. Minai
University of Cincinnati
iyerlr@email.uc.edu
Ali.Minai@uc.edu

Simona Doboli and Vincent R. Brown
Hofstra University
Simona.Doboli@hofstra.edu
Vincent.R.Brown@hofstra.edu

It is generally believed that cognition involves the self-organization of coherent dynamic functional networks across several brain regions in response to incoming stimulus and internal modulation. These context-dependent networks arise continually from the spatiotemporally multi-scale structural substrate of the brain configured by evolution, development and previous experience, persisting for 100-200 ms and generating responses such as imagery, recall and motor action. In the current paper, we show that a system of interacting modular attractor networks can use a selective mechanism for assembling functional networks from the modular substrate. We use the approach to develop a model of idea-generation in the brain. Ideas are modeled as combinations of concepts organized in a recurrent network that reflects previous associations between them. The dynamics of this network, resulting in the transient co-activation of concept groups, is seen as a search through the space of ideas, and attractor dynamics is used to "shape" this search. The process is required to encompass both rapid retrieval of old ideas in familiar contexts and efficient search for novel ones in unfamiliar situations (or during brainstorming). The inclusion of an adaptive modulatory mechanism allows the network to balance the competing requirements of exploiting previous learning and exploring new possibilities as needed in different contexts.

1 Introduction

A consensus is gradually developing that cognition involves the continual self-organization and dissipation of functional networks across several brain regions – especially the neocortex – in response to incoming stimulus and internal modulation [25, 19, 18, 21, 23, 42, 3, 37]. Each functional network emerges from the brain's substrate in response to contextual information, persists while the context applies, and then dissolves back into the substrate to allow a new network – or networks – to emerge.

As pointed out by Doyle and colleagues [8, 7], useful systems are typically *heterogeneous* and *specific* rather than homogeneous and generic. The configuration of such specific heterogeneity usually requires optimization, but that is not feasible in real-time for a complex system like the brain. Instead, such systems must work through self-organization arising naturally from the structure and dynamics of the system. However, rapid self-organization of functional networks is only possible in the brain if it provides the structure and mechanisms that facilitate the process – and it does! The cortex is organized into modules called *cortical columns* that group together into larger modules termed *hypercolumns* [28]. Such modularity is a fundamental enabling mechanism for self-organized complexity in living systems [38], and provides exactly the sort of flexibility that is needed for efficient reconfiguration of functional networks.

We postulate that five factors combine to produce the emergence of effective, flexible, robust and reliable functional networks in the brain. These are:

1. *A modular substrate with sufficient diversity:* The underlying network, which is configured over the multiple time-scales of evolution, development and experiential learning, provides modules with a wide variety of functional micro-behaviors.

2. *A dynamic selective process to bind functional structures:* This process selectively combines an appropriate set of modules so that the correct macro-functionality emerges from the interaction of their micro-behaviors [4].

3. *A dynamic modulatory process to control scope and switching:* This process modulates the excitability of neural units to determine which ones participate in the current functional network, thus controlling the *effective breadth* of these networks, and the transition between networks.

4. *An evaluative feedback process:* This provides a reinforcement signal back to the system so that it can appropriate functional networks can be configured and triggered.

5. *A repertoire of learning processes:* These include: a) Self-organization of micro-behaviors in modules to provide a good behavioral basis; b) Reinforcement-driven Hebbian learning to associate contexts with appropriate functional networks; c) Reinforcement-driven Hebbian learning to

configure the interactions among modules so that useful functional networks become embedded in the substrate through self-organization.

In order to study these principles in a concrete, albeit much simplified, framework, we consider a model for idea generation in a neural system.

A major motivation for the development of the model presented here is our goal to develop a detailed, neurally plausible model of the process of creative idea generation commonly called 'brainstorming'. Brainstorming refers to idea generation under specific guidelines designed to promote quantity and creativity with minimal censorship and criticism [32]. Although these guidelines were designed for use in groups, individuals can obviously engage in creative idea generation as well. In fact the vast majority of laboratory research on brainstorming finds that an equal number of solitary brainstormers outperform interactive groups by almost a 2 to 1 margin when quantity of ideas is counted [13, 29, 35]. Nonetheless being exposed to the ideas of others can be stimulating and, under conditions designed to reduce social inhibition, groups can match or exceed the performance of an equal number of solitary brainstormers, thus closing or eliminating the 'group productivity gap' [33, 36]. Theoretically groups provide the stimulation necessary to get individuals "out of a rut" by activating less accessible categories of ideas and activating atypical sequences of ideas that can be fuel for novel conceptual combinations [6, 30]. There is accumulating evidence for the stimulating effects of the exposure to others' ideas [10, 16, 26] (although see Nijstad et al. [31] for evidence that exposure to other ideas can also have interfering effects on brainstorming). Recent models based on brainstorming as activation, search, and recall of ideas in associative memory can account for a number of these empirical results and have proven fruitful in generating testable hypothesis about a number of important cognitive processes involved in brainstorming, including attention, working memory, memory accessibility, and convergent vs. divergent thinking [6, 34, 30]. One major limitation of these models is the inability to account for the important process of conceptual combination in generating creative responses. One goal of the model presented here is to provide a neurocomputational mechanism for the generation of novel conceptual combinations. In addition, with recent work on the neuroscience of creativity [22, 24, 15], there is the need for more detailed neurally-inspired models of the brainstorming process. Prototypes of the model presented here based on the attractor network architecture have shown promise in accounting for some of the basic empirical brainstorming results in both individuals and groups [5, 14].

2 Problem Formulation

We begin by postulating that *ideas* are combinations of *concepts*, which are the basic representational units in our model. Novel ideas are conceptual combinations whose elements have not been combined in the past, while groups that have been formed previously represent familiar ideas. A cognitive system must be able to retrieve familiar ideas and to generate novel ones.

Ideas always arise – and make sense – within a *context*, e.g., a task situation. Each context – even if it is not completely familiar – tends to elicit a set of ideas by association and suppresses others, presumably making the search for useful ideas more efficient. Thus, a context, Φ, can be seen as a semantic biasing mechanism that preferentially *implicitly* unmasks a subset, I^{Φ}, of ideas from the space of all ideas, allowing a *search process* to explore this subset and "discover" good ideas. Since ideas are combinations of concepts, we model this process as the selective activation of a *context-dependent concept network (CCN)* whose states represent ideas, and whose dynamics embodies the search process. The CCN is a functional network.

Ideas can be seen more broadly as internal responses of the cognitive system, and idea-generation is essentially no different from the generation of motor responses, memory recall or mental imagery (indeed, mental images can be regarded as ideas too!) It is well known that mental responses fall into two broad categories: Automatic and effortful [12]. The former (automatic) are faster, stereotypical and fluent, while the latter (effortful) appear to entail some type of constructive process or search. It has been proposed that effortful responses require the involvement of working memory [2] or global workspace [1], which functions essentially as a temporary "hidden layer" facilitating direct linkage between stimulus and response. Once the direct linkage is consolidated, the mediating process is cut out, leaving behind an automatic response for the future [12]. Alternatively, working memory may remain involved in complex, context-dependent tasks — again as a biasing hidden layer [27, 17, 11]. While these abstract formulations clearly capture important elements of the phenomenology of response generation, they represent an implicitly teleological view, where automatization of response is seen as an end towards which the mechanism of learning works. In contrast, we seek a neural system where the emergence of novel responses and facilitation of familiar ones *both* arise naturally from the intrinsic dynamics of the system. Embedded in an environment, such a system continually creates responses to the information flowing through it — perpetually adjusting its internal constraints to facilitate especially useful responses. Behavior, in this view, is not a goal or a purpose of the system, but a property.

In the case of idea generation, we develop a simple neural model showing how modular organization, selectivity, modulation, evaluative feedback and reinforcement-driven adaptation combine to produce an efficient search process through the system's natural dynamics.

3 The Idea Generation Process

Search is useful only if it is efficient. Exhaustive search, while guaranteed to succeed, is typically not an option. Thus, our primary focus is on how the search for ideas is *shaped* and *guided* by the system's dynamics so that it is as broad as necessary and as narrow as possible. We only consider this criterion heuristically, and will refine it further in future studies.

Efficiency in search requires effective use of information. There are three

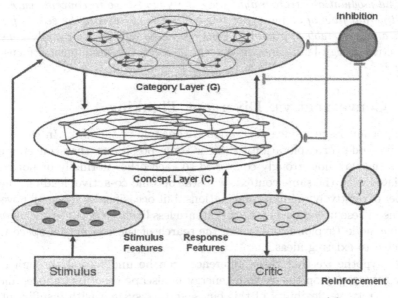

Figure 1: Architecture of the idea generation model.

major sources of information available to guide the search for ideas in our model, and form the basis of its utility. These are:

1. *Current task or domain context:* The search should focus on concepts that, in combination, are likely to be relevant to the context at hand. A neural system does this by associating context representations with specific patterns of module activation, forming functional networks.

2. *Previous experience:* The flow of the search should be guided, and its extent determined, by the experience of the system in inferring the regularities of its operating environment. The only way for a neural system to do this is by embedding such experiential knowledge in its structure and dynamical parameters, so that the emergent flow of the system's activity is appropriately constrained to productive regions and trajectories in the search space.

3. *The progress of the search so far:* As the search proceeds, it should be guided continually by the incoming evaluative feedback towards more productive regions. This is necessary because novel ideas often require the formation of functional networks other than those triggered initially by the context. This can only happen if feedback can overcome the initial bias and create new structures in a systematic way – again, guided by the experiential knowledge implicit in the system's structure.

The primary hypothesis behind our model can be stated in two parts: *1)*

The interaction between afferent context/stimulus information, network structure and modulation create a dynamic energy landscape in concept space; and 2) The itinerant flow of activity over this landscape represents the search for ideas, which are metastable attractors. Thus, the mechanisms for rapidly generating productive energy landscapes in concept space is the fundamental focus of the research.

3.1 Convergent vs. Divergent Thinking

An important issue in idea generation is the style of thinking. In the *divergent thinking* mode (or exploration), there is some possibility that concepts or categories that are not strongly connected to each other normally or not strongly associated with the same context, will still become co-active, leading to a large number of relatively random combinations, but occasionally to useful novel associations. These useful novel associations are less likely to occur in the *convergent thinking* mode (exploitation) where the search of the conceptual space is more restricted to existing ideas.

We hypothesize that these differences can be understood through the dynamics of activity on the dynamic energy landscape described above, and arise from the relative flexibility of this landscape. A system with insufficient modulation and short-term learning is only able to produce stereotypical energy landscapes with relatively high barriers between attractors. It, therefore, tends to get trapped in suboptimal regions of search space and leads to convergent thinking. In contrast, a more flexible system can adapt the energy landscape to create new attractors through the recombination of old ones, thus evincing divergent thinking.

The model we develop embodies this view of idea generator, implementing it in a connectionist framework.

4 Model Description

The model we propose is shown in Figure 1. It comprises the following components:

- A *stimulus* or *input layer*, I, providing a set of n_I *stimulus feature units* (SFUs) encoding afferent stimuli, including context information.

- A *response layer*, R, comprising a fixed pool of n_R *response feature units* (RFUs) denoting response-relevant characteristics of concepts.

- A *concept layer*, C, comprising a pool of n_C *concept units* (CUs). Each concept unit has connections (both excitatory and inhibitory) with a subset of RFUs, thus implicitly defining the response semantics of the concept. The concept layer also has fast recurrent excitatory connections among concept units, which tend to stabilize specific patterns of coactive units, and slower recurrent self-inhibition that *stochastically* limits how long a unit can remain continuously active. These competing tendencies generate a flow of activity patterns — similar to itinerant dynamics over a set of metastable attractors [43] — consistent with the hypothesis that mental constructs are represented as slightly persistent patterns of coherent cortical activity [20, 41]. The patterns of coactive concepts temporarily stabilized by fast connections represent ideas. Each idea activates the response features corresponding to its constituent concepts, which is the effective response of the system. As a whole, the concept layer can be seen as comprising a distributed *semantic network*.

The pattern of recurrent connectivity within the concept layer reflects the utility of associations between concepts based on previous experience. Thus, if CUs i and j have been coactive in several "good" ideas, they have strong positive reciprocal connections, while the connection may be weak, non-existent or even negative if this is not the case. Thus, given the noisy activation and refractoriness of concept units, the dynamics of the concept layer tends to stochastically reactivate attractors that are previously seen good ideas or their mixtures.

- A *category layer*, G, comprising a set of N_S *concept group units* (CGUs) organized into n_Q *cliques*, $\{Q^k\}$, $k = 1, ..., n_Q$. These cliques corresponding (roughly) to utilitarian categories. CGUs within the same clique are (relatively) densely interconnected by excitatory connections, while pairs of cliques are symmetrically connected by relatively sparse excitatory or inhibitory connections *in a specific pattern* reflecting the system's experience of whether joint activity by clique pairs is useful or otherwise. Each CGU has reciprocal connections with a subset of concept units, thus defining a *conceptual group* (CG) which can be seen roughly as "basis functions" for ideas. Several cliques may include the same CG — each associated with its own CGU, and the concepts within a conceptual group may be quite different in terms of their features. The presence of two CGs in the same clique indicate that they have been useful together *in the same context* at various times, implying that good ideas can be elicited in similar contexts simply by activating CGs in this clique. This utilitarian clustering typically means that cliques are also semantically distinct from each other, so that every clique does not include concepts with all possible features. Thus, while the concepts and CGs associated with a clique, Q^k, are quite heterogeneous, they cover only a subset, R^k, of response features, and only a subset in the space of possible CGs.

It should be noted that the modular structure imposed by layer G on layer C is consistent with but *not* identical to the structure implicit in the recurrent connectivity of C. Concepts may be strongly connected even if they do not share a clique if they have been part of good ideas in the past. Over time, learning should adjust clique memberships and conceptual groups to remove this "mismatch", but the current model does not address that issue yet.

- A critic, Ψ, that takes input from layer R and compares the activity pattern of this layer with an internally stored *criterion pattern* of features, generating a graded scalar *response evaluation*. This response evaluation is used as feedback by the system to modulate its parameters, and to control learning. The critic is intended to be a *phenomenological* model of internal and external evaluative mechanisms (e.g., a dopamine signal triggered by a reinforcement signal [40, 39]).

Afferent input from the stimulus layer drives all three layers through modifiable connections. The category layer is also subject to adaptive inhibitory modulation, which controls the total amount of activity allowed in the layer. Competitive inhibition in the concept layer also constrains the number of simultaneously active concept units.

5 System Functionality

As discussed earlier, the system's response is generated through a transiently stable (metastable) spatiotemporal pattern of activity spanning the whole system — a reverberatory (or resonant) [9] pattern involving the category, concept and response layers. However, it is specifically the activity of the concept layer that represents the system's internal response at the level of ideas. This response is projected into the common semantic space spanned by the response features, where it can be evaluated. These features can thus be seen as "internal actuators", or as verbalization components. Given a context (and possibly a stimulus stream), the goal is to generate responses that meet the functional criteria known to the critic as efficiently, rapidly, and copiously as possible.

Processing starts when the system is stimulated by a *context stimulus*, Φ^j through layer S. This results in the activation of one or more cliques in G based on the association between the stimulus pattern and CGUs, and creates a stable activation pattern in layer G that persists even after the context stimulus is removed. This G-layer activity projects a selective bias onto Layer C, creating an implicit energy landscape in concept space with ideas as attractors. However, since only a small number of CUs can be co-active and individual CUs can remain active only for limited durations, the energy landscape keeps changing as ideas emerge, persist for a brief time, and dissolve. This "sticky" flow of ideas in C is the search process shaped by the interaction of bias from G and the recurrent connectivity in C.

As ideas are produced, they generate a stream of evaluations, $\phi(t)$, through the critic. This is integrated by a low-pass filter

$$\bar{\phi}(t) = \alpha\phi(t) + (1 - \alpha)\bar{\phi}(t - 1); \quad 0 < \alpha < 1$$

and modulates the inhibition level on layer G. As $\bar{\phi}(t)$ becomes lower (because no good responses have been found), inhibition on G is also lowered, eventually leading to the activation of more cliques and their CGs. This changes the bias on Layer C and, therefore, the energy landscape for the flow of ideas. The order in which new cliques are activated as inhibition is lowered is robustly dependent on two factors:

1. The original context stimulus, which determined the original clique activity pattern.

2. The inter-clique connectivity reflecting the expected utility of co-activating certain cliques (or CGUs within cliques).

This broadening of the search for ideas is useful, but is not likely to be very efficient without a concurrent process for narrowing the search so it can focus in more productive regions of idea space. This is achieved using feedback from the concept layer to the category layer. As the search is widened and some good ideas start appearing, this results in a gradual increase in $\bar{\phi}$, which then tries to *decrease* the activity in G. Also, each time a good idea is rewarded sufficiently, two processes are triggered:

1. *Short-term performance-dependent reorganization of G-layer connectivity*: The reward signal causes the connections from the concept layer to the category layer to be transiently, but strongly, potentiated (shown by the gate in Figure 1). Thus, the concept units active at the time of the reward (which presumably comprised the good idea) send an unusually strong excitation to the CGUs active at the same time, which respond to this jolt by firing strongly (e.g., bursting). This, in turn, leads to short-term synaptic potentiation of connections between bursting units [18], and a short-term depression of connections between bursting and non-bursting units. The net result is that Layer G units projecting to concepts in good ideas become relatively more strongly connected to each other, as do the cliques in which they reside, *thus changing the G-layer connectivity to reflect recent experience.*

2. *Permanent performance-based change in association between concepts*: The reward signal causes the concept units active at the time of the reward to permanently strengthen their excitatory connectivity.

The effect of (2) is that good ideas become more embedded in the concept layer, which makes them likelier to arise in the future. The effect of (1) is more pertinent to the current search. The change in G-layer connectivity means that the cliques likely to survive the process of increasing inhibition due to greater reward are not necessarily those originally activated by the context, but those responsible for the recent good ideas. If more good ideas are forthcoming from this restricted set of cliques, the system will go on restricting the search further. If the process becomes too restrictive (and good ideas cease), inhibition will drop again and the search will widen. However, the pattern of widening will be affected by the recent experience of the system because the original inter-clique connectivity has been modified (transiently) by this experience. Eventually, the system will reach a stage where a set of cliques dense in good ideas is activated, so that the increased reinforcement sustains this set even at maximum inhibition. Basically, the system has arrived at the best set of cliques – and thus the best pool of ideas – through a process of "intelligent annealing" that takes into account the current context, the system's past experience, and the results of the system's current search. This discovered set of cliques and/or concepts/ideas can be associated with the original context, and slightly change the inter-clique connectivity to reflect the new discovery. More significant changes such as realignment of clique membership are left for the future.

6 Results

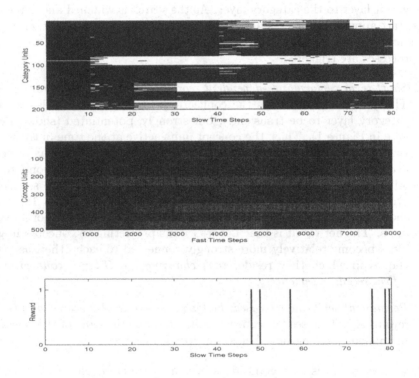

Figure 2: Simulation of the search process.

A somewhat simplified version of the model described above has been implemented, and search processes simulated using this model. The main simplification in the simulations is that each concept is typically associated with CGs from only one or two cliques. The cliques themselves are connected but nonoverlapping (i.e., they do not share CGs), and both category and concept layer units are arranged such that units associated with the same clique are plotted together in figures. The dynamics of the simulated network incorporates two

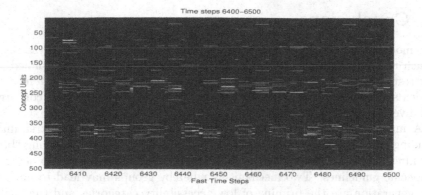

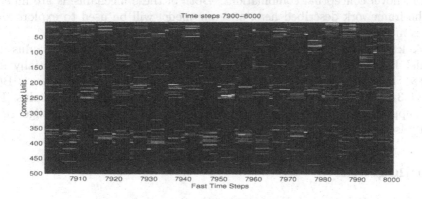

Figure 3: Simulation of the search process.

time-scales: A slow scale over which the CG activations are updated, and a fast scale used to update the concept units. Thus, each cycle of G layer update embeds 100 steps of the fast cycle. Each concept unit, however, can only be active for a random duration between 4 and 8 fast steps, and must then be refractory for a similar random duration. This allows for the context-dependent itinerant dynamics to emerge in concept space.

Figure 2 shows an example run of the simulator. The G layer has 10 cliques of 20 units each, while the C layer has 500 concept units. The system emulates a situation where good ideas arise preferentially from concepts associated with cliques 3, 6 and 10 *together*. The search starts by activating clique 6, and gradually discovers the right combination of cliques through the process described above. At that point, rewards become much more dense in time, indicating success.

Figure 3 shows a close-up view of the concept layer dynamics at different stages of the search process. The itinerant dynamics across idea space can be seen clearly.

7 Conclusions

The model described in this paper embodies a general, complex systems approach to idea generation and cognitive response – albeit in a simplified way. Our results show that the model is capable of performing an intelligent search in idea space through its inherent dynamics, and can represent the convergent and divergent modes of thinking.

A major motivations for the current model is to provide a better understanding of the cognitive, computational and neural processes underlying "brainstorming", i.e., idea generation by individuals interacting in a group according to specific guidelines. Two mechanisms whereby groups may lead to improved idea generation are the priming of low-accessibility categories and the facilitation of novel conceptual combinations. Both of these mechanisms are inherent in the framework described here, and our model will be used to explore these issues in a systematic fashion.

Acknowledgements: The authors wish to thank Paul Paulus and Daniel Levine for useful ideas and insights. This work was partially supported by the National Science Foundation through award nos. BCS-0728413/0729470/0729305 to Ali Minai, Vincent Brown and Paul Paulus. Partial support also came from the Deputy Director of National Intelligence for Analysis.

Bibliography

[1] BAARS, B.J., and S. FRANKLIN, "How conscious experience and working memory interact", *Trends in Cognitive Sciences* **7** (2003), 166–172.

[2] BADDELEY, A., *Working Memory*, Oxford University Press, New York (1986).

[3] BRESSLER, S.L., and E. TOGNOLI, "Operational principles of neurocognitive networks", *International Journal of Psychophysiology* **60** (2006), 139–148.

[4] BROOKS, R.A., *Cambrian Intelligence: The Early History of the New AI*, MIT Press (1999).

[5] BROWN, V., and S. DOBOLI, "A neural network simulation of interactive group brainstorming", *Eighteenth Annual Convention of the Association for Psychological Science*.

[6] BROWN, V., M. TUMEO, T. LAREY, and P. PAULUS, "Modeling cognitive interactions during group brainstorming", *Small Group Research* **29** (1998), 495–526.

[7] CARLSON, J.M., and J. DOYLE, "Highly optimized tolerance: Robustness and design in complex systems", *Physical Review Letters* **84** (2000), 2529–2532.

[8] CARLSON, J.M., and J. DOYLE, "Complexity and robustness", *Proceedings of the National Academy of Sciences USA* **99 Suppl. 1** (2002), 2539–2545.

[9] CARPENTER, G.A., and S. GROSSBERG, "A massively parallel architecture for a self-organizing neural pattern recognition machine", *Computer Vision, Graphics, and Image Processing* **37** (1987), 54–115.

[10] COSKUN, H., PAULUS, V. P.B., BROWN, and J.J. SHERWOOD, "Cognitive stimulation and problem presentation in idea generation groups", *Group Dynamics: Theory, Research, and Practice* **4** (2000), 307–329.

[11] DECO, G, and E.T. ROLLS, "Attention, short-term memory, and action selection: A unifying theory", *Progress in Neurobiology* **76** (2005), 236–256.

[12] DEHAENE, S., M. KERSZBERG, and J.-P. CHANGEUX, "A neuronal model of a global workspace in effortful cognitive tasks", *Proceedings of the National Academy of Sciences, USA* **95** (1998), 14529–14534.

[13] DIEHL, M., and W. STROEBE, "Productivity loss in brainstorming groups: Toward the solution of a riddle", *Journal of Personality and Social Psychology* **53** (1987), 497–509.

[14] DOBOLI, S., A.A. MINAI, and V.R. BROWN, "Adaptive dynamic modularity in a connectionist model of context-dependent idea generation", *Proceedings of the International Joint Conference on Neural Networks*, (2007).

[15] DUCH, W., "Intuition, insight, imagination and creativity", *IEEE Computational Intelligence Magazine* (August 2007), 40–52.

[16] DUGOSH, K.L., and P.B. PAULUS, "Cognitive and social comparison processes in brainstorming", *Journal of Experimental Social Psychology* **41** (2005), 313–320.

[17] DUNCAN, J., "An adaptive coding model of neural function in prefrontal cortex", *Nature Reviews: Neuroscience* **2** (2001), 820–829.

[18] EDELMAN, G.M., and G. TONONI, *A Universe of Consciousness: How Matter Becomes Imagination*, Basic Books (2000).

[19] FREEMAN, W.J., *How Brains Make Up Their Minds*, Columbia University Press (2000).

[20] FREEMAN, W.J., "Cortical aperiodic shutter enabling phase transitions at theta rates", *Proceedings of the International Joint Conference on Neural Networks*, (2007).

[21] FUSTER, J.M., *Cortex and Mind: Unifying Cognition*, Oxford University Press (2003).

[22] HEILMAN, K.M., S.E. NADEAU, and D.O. BEVERSDORF, "Creative innovation: Possible brain mechanisms", *Neurocase* **9** (2003), 369–379.

[23] IZHIKEVICH, E.M., *Dynamical Systems in Neuroscience: The Geometry of Excitability and Bursting*, MIT Press (2007).

[24] JUNG-BEEMAN, M., E.M. BOWDEN, J. HEBERMAN, J.L. FRYMAIRE, S. ARAMBEL-LIU, R. GREENBLATT, P.J. REBER, and J. KOUNIOS, "Neural activity when people solve verbal problems with insight", *PLoS Biology* **2** (2004), 500–510.

[25] KELSO, J.A.S., *Dynamic Patterns: The Self-Organization of Brain and Behavior*, Cambridge, MA: MIT Press (1995).

[26] LEGGETT, K.L., "The effectiveness of categorical priming in brainstorming" (1997).

[27] MILLER, E.K., "The prefrontal cortex and cognitive control", *Nature Reviews: Neuroscience* **1** (2000), 59–65.

[28] MOUNTCASTLE, V.B., "The columnar organization of the neocortex", *Brain* **120** (1997), 701–722.

[29] MULLEN, B., C. JOHNSON, and E. SALAS, "Productivity loss in brainstorming groups: A meta-analytic integration", *Basic and Applied Social Psychology* **12** (1991), 3–23.

[30] NIJSTAD, B.A., and W. STROEBE, "How the group affects the mind: A cognitive model of idea generation in groups", *Personality and Social Psychology Review* **3** (2006), 186–213.

[31] NIJSTAD, B.A., W. STROEBE, and H.F.M. LODEWIJKX, "Cognitive stimulation and interference in groups: Exposure effects in an idea generating task", *Journal of Experimental Social Psychology* **38** (2002), 535–544.

[32] OSBORN, A.F., *Applied Imagination*, New York: Scribner's (1957).

[33] PAULUS, P., and V. BROWN, "Dennis, a.r. and williams, m.l.", *Group Creativity*, (P. PAULUS AND B. NIJSTAD eds.). Oxford University Press New York (2003), pp. 160–178.

[34] PAULUS, P., and V. BROWN, "Enhancing ideational creativity in groups: Lessons from research on brainstorming", *Group Creativity*, (P. PAULUS AND B. NIJSTAD eds.). Oxford University Press New York (2003), pp. 110–136.

[35] PAULUS, P.B., and M.T. DZINDOLET, "Social influence processes in group brainstorming", *Journal of Personality and Social Psychology* **64** (1993), 575–586.

[36] PAULUS, P.B., and H.C. YANG, "Idea generation in groups: A basis for creativity in organizations", *Organizational Behavior and Human Decision Processes* **82** (2000), 76–87.

[37] PERLOVSKY, L.I., and R. (eds.) KOZMA, *Neurodynamics of Cognition and Consciousness*, Spinger (2007).

[38] SCHLOSSER, G., and G.P. WAGNER (EDS.), *Modularity in Development and Evolution*, University of Chicago Press, Chicago, IL (2004).

[39] SCHULZ, W., "Multiple reward signals in the brain", *Nature Reviews Neuroscience* **1** (2000), 199–207.

[40] SCHULZ, W., P. DAYAN, and P.R. MONTAGUE, "A neural substrate of prediction and reward", *Science* **275** (1997), 1593–1599.

[41] SPIVEY, M., *The Continuity of Mind*, Oxford University Press (2007).

[42] TONONI, G., "An information integration theory of consciousness", *BMC Neuroscience* **5** (2004), 42.

[43] TSUDA, I., "Towards an interpretation of dynamic neural activity in terms of chaotic dynamical systems", *Behavioral and Brain Sciences* **24** (2001), 793–847.

Chapter 17

An Introduction to Complex-System Engineering

Michael L. Kuras

© 2007 Michael L. Kuras

INTRODUCTION

This is an introduction to complex-system engineering (cSE). cSE is going to become the second branch of system engineering. It is still in its formative stages. All of cSE's presently known ideas are briefly discussed. A familiarity with the first branch of system engineering, so called traditional system engineering (TSE), is assumed.

MOTIVATION AND OVERVIEW

TSE is not applicable to every problem. Its applicability is limited by the assumptions on which it rests. TSE's assumptions are essentially the assumptions of **reductionism**. Reductionism, in its essence, is the belief that *any* portion of reality – including all of it – can be understood or comprehended by understanding the parts of that reality and composing a mental model of the greater reality exclusively from those parts. All of the parts of some portion of reality, and the relationships among those parts, can directly account for *everything* that can be known or conceptualized about the greater reality. The parts of such a whole can also be resolved into parts and understood in the same manner. This isn't true all of the time.

To reject the universality of reductionism is to accept that there are properties that can only be associated with the whole of something – but not with its parts. Such properties are frequently termed **emergent**. In addition, when the relationships or interactions among the parts of a whole are considered, it is not always possible to find a way to consider them sequentially that is equivalent to their consideration in parallel. Even if system engineers are willing today to accept these non reductionist propositions, they do not know how to transfer that acceptance into the practice of their discipline.

Much of the debate about reductionism versus its alternative, **holism**, is cast in terms of reality itself. This is a serious mistake. It overlooks the role that the human brain might play. It is the human brain that composes the mental models that are our only window on reality. The human brain is finite. The consequences of this simple fact are almost always overlooked.

Accounting for the consequences of a finite human brain leads to multi scale analysis (MSA). Multi scale analysis is one of the key ideas in complex-system engineering. Validating the need for multi scale analysis is exceedingly difficult. This is because any conceptualization of reality (a mental model of that reality) can never be directly compared with reality itself. Validation must be indirect.

The second major idea in complex-system engineering is evolution. Evolution is also inconsistent with reductionism. Most system engineers are unfamiliar with evolution. Most have heard of it – but they are not familiar with it in any depth, and not with the processes that account for it. And certainly system engineers do not know how to connect evolution to the engineering of systems.

Evolution is frequently associated exclusively with biology. This is too restrictive. Evolution is much more widespread than that. Evolution is also context dependent. The processes that drive evolution vary as that context changes. However, whatever the processes might be in any specific context, they always express in some fashion what are termed the predicates of evolution. These will be identified, as well as what can be considered an explanation for evolution in any context. Both of these topics are important in complex-system engineering.

Given multi scale analysis and evolution, there is one more key idea in complex-system engineering. Because cSE is a branch of system engineering, it must have a set of methods. Engineering, including system engineering, is always about using a set of methods to establish an equivalence between a real world problem and a real world solution. The methods in cSE are termed the Regimen of complex-system engineering (Kuras, 2004). They are just listed here for now. Their context sensitive formulation and application depend upon multi scale analysis and evolution.

THE REGIMEN OF COMPLEX-SYSTEM ENGINEERING

1. Define Outcome Spaces.

2. Selectively perturb self-directed development.

3. Characterize Continuously.

4. Perturb the environment.

5. Specify rewards and penalties for autonomous agents.

6. Judge outcomes and allocate prizes, accounting for side effects.

7. Formulate and apply stimulants for synergistic self-directed development.

8. Formulate and apply safety regulations.

MULTI SCALE ANALYSIS

In order to appreciate the limits of reductionism and how to get beyond them requires an examination of how engineers (in fact, all humans) think – about how they compose problems and solutions in their heads, or how they build mental models of the real world. This should quickly become a question about how the human brain works. This may seem a bit daunting to the average engineer. But it has to be addressed. How does the human brain form and manipulate conceptualizations of reality? And does this influence what an engineer thinks of as real problems and real solutions? Are there built-in characteristics of the human brain that prevent engineers from having a completely "transparent" window on reality? In other words, does the brain alter or transform or filter in some way what an engineer assumes to be reality as he or she builds and manipulates mental models of that reality?

The answer is that the way that the human brain works does influence what is comprehended as reality. The window on reality that the brain provides is not a completely transparent one. The human comprehension of reality is not distortion free. What makes this very difficult to deal with is that it is not possible to comprehend anything without that influence being present. Nothing can be understood without conceptualizing it. And it is not possible to directly make "before and after" comparisons of the brain's influence on conceptualizations of reality. Nonetheless, the influence of the brain on what is conceptualized should not be ignored. Multi scale analysis explicitly acknowledges this influence.

Human conceptualization is limited. This is because **the human brain is finite**. Its capabilities and capacity are both bounded. It is this limitation on capability and capacity that produces a less than transparent window on reality. Surprisingly, this characteristic has largely been ignored in the prior work on this issue. Maybe it is too obvious. But for now, here are some of the consequences. All of reality cannot be conceptualized at once. (We can't even conceptualize all that we think that we know at once.) And there is no way to ensure that combining partial conceptualizations of reality (either in their entireties or in some abstracted form) will yield a comprehensive conceptualization of the greater totality. Conceptualizations of reality, mental models, are *always* partial – and limited to one "scale" at a time. The human brain can, however, change its scale of conceptualization altering what is included and omitted in mental models of reality.

These are statements about how the human brain works and nothing more. For example, all of the properties and behaviors of all of the particulates of a cloud, and all of the properties and behaviors of the cloud as a whole cannot be conceptualized at the same time. It is not possible to conceptualize water and all of its properties while conceptualizing all of the individual atoms of Hydrogen and Oxygen that comprise the water. It is not possible to simultaneously conceptualize, say, the full meanings of individual letters or words and the meanings of whole sentences and paragraphs. It is not possible to simultaneously conceptualize all of the individual bees of a beehive and the totality of that beehive. The human brain doesn't work that way. It generates separate mental models for these aspects of the same reality; and it can only toggle between these mental models. Conceptualizations or mental models are based on the brain's internally developed frames of reference. And the brain can and does employ different frames of reference as it forms mental models of different aspects of the same reality.

Frames of reference do not refer just to the notions of space and time. They are the basis for *every* pattern that is a portion of every mental model of reality that the human brain generates: for example of color, of taste, of ownership, of authority, of matter or substance, of quality and quantity, and so on. And sometimes the frames of reference used in mental models of the same reality are not just different but incompatible as well. A change in the scale of conceptualization occurs when the brain shifts between incompatible frames of reference.

The entire content of every conceptualization (of every mental model that the brain might form) is exactly and only a finite set of patterns; memories are basically stored versions of earlier conceptualizations; the capacity for storing patterns is also finite or limited; every current conceptualization depends on earlier conceptualizations (i.e., draws on memory); and frames of reference are the deepest and most condensed forms of memory. These points elaborate what it means to say that the human brain is finite, and that it is limited or bounded in its capability to form and retain conceptualizations of reality. If these points were further elaborated, the consequences just outlined would follow.

What is crucial here, however, is that mental models of reality, based on incompatible frames of reference, simply do not combine as suggested by reductionism. And any analysis, based solely on such premises, is going to be partial and even flawed.

Multi scale analysis allows for, but does not rely exclusively on, reductionist techniques. It uses, in addition, **statistical analysis** to find and confirm relationships that span scales of conceptualization – that is, between mental models of the same reality that are based on incompatible frames of reference. Even at a single scale of conceptualization, multi scale analysis departs from an exclusively reductionist approach. That departure is **regime analysis**, which is based, at least, on Radial Category Theory and Reed network models. Regimes are non-disjoint and non-hierarchical partitionings. Radial Category Theory came out of Linguistics, and Linguistics is a branch of Cognitive Science – which focuses on how we think and communicate. Reed networks are discussed briefly below.

In summary, multi scale analysis explicitly acknowledges that many of our mental models of reality cannot be combined or related to one another in a purely reductionist fashion. At present the only way to find or to explore cross scale relationships is with statistical analysis. Moreover, at any one scale of conceptualization, reductionist analysis can be augmented with regime analysis. The presently known techniques applicable to regimes are based on Radial Category Theory and Reed networks.

EVOLUTION

Evolution is the next major idea. Evolution is a label for distinguishable differences in successive generations of a population.

Evolution is a label for *changes* that are seen as gradual or progressive or cumulative but *not* repetitive (like the changing seasons or a pendulum), and that are not viewed as arbitrary or random. Further, evolutionary changes are not the direct consequence of an explicitly identifiable outside agent's intervention (such as a constantly applied force) – and so are frequently termed *self-directed*. Such changes are generally understood to apply to the substance and structure of things as well as to their behavior. Since this form of change is not seen as arbitrary, it may be due to some process or processes. *Theories of evolution* are attempts to identify and characterize these processes. Theories of biological evolution are the ones most frequently and explicitly examined. In fact, most people still associate evolution exclusively with biology.

Evolution deals with self-directed changes in populations. What this means for a system engineer is that a system must be treated as a set of populations. Evolution unfolds at multiple scales of conceptualization. And it is always context dependent. Said another way, the evolution of a system – that is, self-directed changes in a set of populations – can never be wholly disconnected from the environment of those populations. Both the populations and their environment must be considered and analyzed if the system is to be engineered. It simply is not possible to disjointly partition an evolutionary system from its environment.

Evolution is driven by a set of evolutionary predicates. Typically, five or fewer are identified. But this can be misleading. These predicates will always be expressed as context specific processes. These processes are continually generating outcomes (as modifications or changes in the populations of the system). The Regimen of cSE does not directly seek to produce such outcomes itself. Instead, the Regimen's methods operate on these processes.

These are the five commonly identified predicates of evolution.

1. **Member Addition**: The size of a population increases (through reproduction, through recruiting, etc.).
2. **Similarity** (Heredity): New members are similar to existing members. (In the specific case of reproductive increase, the characteristics of new members are directly derived from the characteristics of existing members.)
3. **Variation**: The characteristics of new members are not uniform.
4. **Adaptation**: Members of a population do not behave uniformly and behavior is not independent of other members and the environment.
5. **Selection**: Members of a population are subject to attrition as a consequence of their characteristics and behavior.

Generally, the rate of new member addition will be greater than the rate of member attrition. This is usually referred to, in biological contexts, as super fecundity. However, there can be temporary exceptions.

There are actually more predicates, although they are frequently overlooked. Not all populations evolve. And not all *finite* **populations** evolve. The necessary and sufficient condition for this is identified below. The five predicates listed above are still necessary, however; but they are not sufficient by themselves. For example, if all members behave uniformly the population will not evolve. At the same time, if any of the other four predicates is not expressed in some way, a population will not evolve *even if* behavior is not uniform. But there is more.

The persistence of a whole population must be much greater than that of any of its members. This is almost always overlooked as well. Roughly speaking, there are **no immortal members** in an evolving population. This is why it is foolish to take too literally the often repeated phrase that "survival of the fittest" is the essence of evolution. It's not. No member of an evolving population, no matter how fit it might be – even the fittest in a population – is going to survive indefinitely. If that were to happen, then the population itself would cease to evolve. But so called fitness is a useful measure in understanding evolution. It is a way to score the persistence of members and populations. And it can be quantified. In general, fitness is always less than 1. An informal definition of fitness is provided below.

Members must also exhibit **life cycles**. A life cycle is the period of existence from inception to elimination through attrition. New members must **mature** before they are full members of an evolving population. The period of maturation may be extremely brief – and even confined to the moments of inception – but it is a necessary facet of evolution. Only mature members can participate in (contribute to) the addition of new members. The processes of maturation are usually distinct from those of evolution, even given specific populations. Maturation recapitulates prior evolution; but recapitulation should not be interpreted as simply repeating prior evolution. The processes of maturation and of evolution do, however, share the same predicates.

There is a necessary and sufficient condition for a system of finite populations to evolve. It is the accumulation of complexity by the system. If this isn't happening, then evolution isn't happening. Roughly speaking, if one system is more evolved than another, then one has accumulated more complexity than the other. This makes an understanding of complexity important in complex-system engineering. Informal definitions of complexity and fitness are provided below – and then used to make an important distinction.

Price's equation is currently the best summary of evolution at a given scale of conceptualization. It is a useful statistical relationship. But it is still a very partial summary of evolution. Much work remains to be done in understanding evolution. That task is by no means complete.

$$\bar{w}\Delta\bar{z} = COV(w_t, z_t) + E(w_t \Delta z_t)$$

Price's equation is applicable to a single scale of conceptualization of an evolving system, but it is not expressed directly in terms of complexity. Instead, it is expressed in terms of individual characteristics or behaviors and their connection to fitness. In Price's equation, w represents fitness, and z a specific characteristic or behavior.

Regardless, the complexity of a system can accumulate across multiple scales of human conceptualization. To understand how that might unfold, and in the absence of a better version of Price's equation, network models can be used.

Finite populations can evolve. When they do, their capacity for accumulating complexity is not unbounded at any one scale of conceptualization (since they are, after all, finite). Nonetheless, the capacity of a system to accumulate complexity is seemingly unbounded. Network models can help to understand this phenomenon.

This can be done by associating members with **nodes** and the organizational and behavioral relationships among members with **edges** or links. The complexity of a population grows as membership increases and the relationships among members increase. There are three types of network models available to capture this. They differ primarily in terms of the characteristics of their edges. Basically edges can be one directional, two directional or without direction. Reed networks have edges that are non directional. Reed network models are not widely known or used. They are an important part of multi scale analysis.

As complexity accumulates in a system, there is a progression in the network types that should predominate in modeling the populations of a system: Sarnoff → Metcalf → Reed → Sarnoff → ... This progression is tied to the incremental costs of adding additional nodes and edges. This cost is measured in the "energy" needed to accomplish the addition. The least costly alternatives available to the system to increase the amount of complexity that it accumulates will persist and eventually predominate.

What is of special note here is the **Reed → Sarnoff progression**. The continued addition of members and relationships in the populations of a complex-system requires an additional scale of conceptualization in order for the added complexity to be humanly comprehended. In short, for a human being to understand what is happening, new frames of reference must be employed in the pertinent mental models. The frames of reference associated with Reed networking in this progression are not appropriate for the added complexity captured by an additional Sarnoff network model at a new scale of comprehension. The Reed network and the additional Sarnoff network use incompatible frames of reference. They cannot be simultaneously comprehended by a human being today. At a minimum the members of the relevant populations will seem to be different.

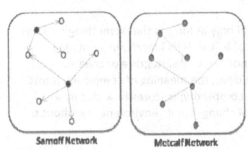

Samoff Network   Metcalf Network

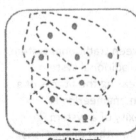

Reed Network

A Sarnoff network uses one-directional edges, represented here as arrows. These edges represent one way flows or relationships from source nodes to sink nodes. Each edge must connect exactly two nodes, but each node can be associated with multiple edges. The second example is that of a simple Metcalf network. It is seen by many as a generalization of the Sarnoff network. There are no distinctions of node types now; they are all the same. As before, each edge connects exactly two nodes. But in this case, flows or relationships can be in one or both directions (from the first node to the second, vice versa, or both). Also as before, a single node can be associated with multiple edges. The third example is that of a simple Reed network, sometimes called a combinatoric network. All of the nodes in a Reed network are notionally the same as with Metcalf networks. In this case, dotted lines represent the individual edges. Individual edges are non directional and *one* edge can interconnect *any* number of nodes, not just two. As before, though, one node can be associated with any number of edges. It is well beyond the scope of this introduction to discuss how to use the various sorts of network models that are available. But it is important to appreciate that they are all available – especially the Reed network model.

In the interests of brevity, only an informal definition of complexity and fitness can be provided here.

Complexity: the measure of a thing's (a system's) available changes *that do not alter its identity*.
Fitness: the measure of a thing's complexity relative to that of the "complexity" of its environment.

Maintaining "identity" is important in understanding both complexity and fitness. Identity is one of the **Modalities of Conceptualization**. These would be introduced in a discussion of a finite human brain and its implications for the formation of mental models or human conceptualization. These Modalities are the facets of how the human brain functions without positing physiological explanations for them – for example in terms of neurological networking or chemistry. That can, of course, be done. Roughly speaking, "identity" is the functionality of the brain that aggregates patterns giving them a cohesion relative to one another that is greater than with all other patterns in a conceptualization.

The complexity of something is meaningful only as long as that something remains cohesive in the human conceptualization of it. This is its identity in a nutshell. Another way of saying this is that it does not die, or disintegrate or collapse. Excessive change can do this, and when it does, the meaning of complexity is lost. Fitness is another measure. It is based on complexity; it scores how closely a system's complexity matches the potential changes in its environment without the system losing its identity.

PULLING IT TOGETHER (PART 1)

Using these definitions, a clear and powerful distinction between traditional system engineering and complex-system engineering can be made. Traditional system engineering is directed at decreasing the complexity and increasing the fitness of a system at a single scale of conceptualization. Complex-system engineering is directed at focusing and accelerating increases in the complexity of a system without a loss of fitness at multiple scales of conceptualization. Other distinctions are also possible. For example, TSE is applicable to problems that can be stabilized, while cSE should only be applied to problems that cannot be stabilized.

And it is now possible to provide an initial definition of a complex-system. A **complex-system** is a system that evolves (self-directed increases of complexity and fitness) at multiple scales of conceptualization. Both this definition and the distinctions between TSE and cSE deserve further discussion, but that is precluded here in the interests of brevity.

A different name is needed for traditional system engineering now. The term traditional suggests orthodox – and so implies that any other approach is unorthodox. And for many unorthodox is very close to wrong, or at least to suspect. The two branches of system engineering should be distinguished according to the kinds of systems to which they are applicable. This is already the case for cSE. TSE should be relabeled reducible-system engineering or rSE.

In order to engineer complex-systems it is best to treat them as populations. When doing so, it is important to be alert to the possibility that a whole population might have properties that cannot be associated with the members of the population. These are its so called emergent properties. However, other member properties (or relationships) might well be related to these emergent properties. This is something that only multi scale analysis will reveal, always with the understanding that statistical correlation is not causation. Also, a whole population at one scale is not always a member of another population at a "higher" scale. This is reductionist thinking. Although this might be true in some cases, it is wrong to assume that this is so in all cases. It's not.

A system exhibits functionality, organization, and substance at one or more scales of conceptualization. Distinct and frequently incompatible frames of reference are necessary for each scale. These frames of reference are not simply those of as space and time, as was noted earlier. Frames of reference are necessary for *every* pattern that our brains assemble as the content of conceptualizations: like color, flavor, ownership, and authority, and not just space and time.

Lastly, in treating a system as a set of populations at multiple scales, a distinction should be made between **autonomic** and **autonomous** members and populations. Primary attention should be focused on the autonomous ones. These are the "agents" that, if present, are most easily associated with the self directed development of the whole system. Individual human beings in a society are examples of autonomous agents. They are driving the evolution of society. A watch and a cell phone are examples of autonomic agents, as is an automobile. Watches and cell phones and automobiles are not driving the evolution of human society – although they may be evolving themselves, *as populations*.

Here are some examples of complex-systems. Brevity, however, precludes their discussion here.

- The Internet; all of the software in the world.
- English (natural languages, generally).
- Ballistic Missile Defense System (not individual missiles, radars, etc.).

- Purpose built fleets of aircraft (not an individual aircraft).
- A human city (not a building or a road).
- The Human Immune System; the Health Care System.
- The Internal Revenue Service (IRS).
- Network Centric Enterprise Infrastructure.
- Sensor swarms.
- A corporation; the national economy; the world economy.
- Religious congregations; a political faction.

Complex-systems evolve. This can also be understood as learning. Whole populations have to learn, even if their members do not. Of course, members can learn too. So complex-system engineering can also be understood as focusing and accelerating the processes of learning. And many of the methods associated with encouraging learning can be transferred to cSE. Terms like "evolutionary engineering" also suggest a focus similar to that of cSE.

REGIMEN OF cSE

The Regimen of cSE is the set of methods that can be used to engineer a complex-system. They were listed earlier; each of them is now discussed very briefly.

The first two methods can be understood as umbrella methods. The other six methods can be understood as more specific elaborations of one or both of these.
- Define OUTCOME SPACES at multiple scales of conceptualization and for multiple regimes. This means thinking globally but not always disjointly and in terms of large sets of acceptable possibilities, not in terms of a lattice or web or hierarchy of specifically required outcomes (properties or behaviors).
- Selectively PERTURB SELF-DIRECTED DEVELOPMENT (organizational and behavioral) at specific scales and in specific regimes. This means intervening locally, without exercising control, and expecting side effects.

Outcome Spaces in cSE may seem roughly similar to establishing requirements in rSE – except that Outcome Spaces are far more general and less persistent than are requirements specifications. And they need not be complete. They can even seem contradictory to some extent. Outcome Spaces broadly identify what are currently viewed as desirable properties, relationships, or interactions for the populations and population members of a complex-system. Perturbing self directed development in cSE may seem roughly similar to taking direct action to control development in rSE. Control is important in distinguishing the methods of cSE from those of rSE. **Control** [of a thing] is the realization of predictable and persistent consequences through actions [on a thing] that supersede or preempt any other actions [imposed on or self-initiated by a thing]. cSE does not seek to control the development of a system; rSE does.

- **CHARACTERIZE CONTINUOUSLY.** This means tracking the populations and population members of a complex-system, noting the changes in their properties, relationships, and interactions, as well as in appropriate measures such as complexity and fitness.

A complex-system operates and develops, and therefore changes, continuously. It never shuts down or is "turned off." That is because a complex-system is one that is evolving. Roughly speaking, to "turn off" a complex-system, would mean to kill it. Its operation and development are self-directed. It is important to keep track of this evolution – especially after efforts are directed at influencing that evolution. Continuous characterization is also where multi scale analysis fits most naturally into the Regimen. The characterizations produced are made available not just to the system engineer but to the autonomous agents of the complex-system.

The continuous characterization of a complex-system can be understood to correspond roughly, in rSE, to specifying in advance what a system is to do or to be (the requirements specifications of rSE) in that many of the same skills are needed. But the application of those skills is very different. For those familiar with architecting, continuous characterization is somewhat akin to developing "as is" architectures – except that the "as is" architectures must be continually updated. And attention to continuous characterization is far more important than that of any "to be" architecture in rSE – which might loosely be associated with a set of Outcome Spaces, except that Outcome Spaces are not persistent and should be updated as the system evolves.

- Temporarily **PERTURB ENVIRONMENTALS** in order to influence the self-directed development of a system.

An updated definition of a system is needed. That has been provided elsewhere (Kuras, 2006). For now, it is assumed that the distinction between a system and its environment is understood. One of the ways that the self-directed development of a system can be influenced is to perturb the system's environment. This perturbation can be localized and brief – or more sustained and pervasive. An example that is frequently used is that of watering a garden. This alters the environment of the garden (the garden is the complex-system in the example). But it leaves to the plants in the garden (some of the autonomous agents of the system) to interact with that temporarily modified environment as they see fit as part of their self-directed development.

- Establish specific **REWARDS AND PENALTIES** for autonomous agents in order to influence their self-directed decision making and development.

- Judge cumulative and collective results, not just specific outcomes, in **ALLOCATING PRIZES**; account for side effects.

These next two methods are most easily discussed together – but they are really separate. Rather than specifying what autonomous agents should be doing, the establishment of rewards and penalties alters the factors that autonomous agents might consider in making and carrying out their own decisions about what to do. NASA has begun to use this method in their effort to engineer a persistent return to the Moon and the human exploration of Mars. This contrasts with what they did earlier in the Apollo Project – an effort that eventually died. Rewards involve the establishment of prizes and the criteria to win them. Judging refers not just to assessing the fulfillment of these criteria, but to examining any side effects that accompany them, and then allocating the prizes accordingly. Of course, if autonomous agents do not recognize a persistent linkage between criteria, judging, and prizes, then the efficacy of these methods will be diluted or worse.

- Formulate and apply **STIMULANTS FOR SYNERGISTIC SELF-DIRECTED DEVELOPMENT**.

This next method involves techniques that will alter the number, or the frequency, or the intensity, or the persistence of interactions or relationships among the autonomous agents of a system – with little or no regard for the details of specific interactions or relationships. This method is very much like stirring the pot. The necessary ingredients are presumed to be in place. This method involves techniques that accelerate – or decelerate – the interactive processes of evolution.

- Formulate and enforce **SAFETY REGULATIONS**.

This last method explicitly recognizes that cSE focuses and accelerates what is otherwise a natural unfolding of development in a complex-system. This acceleration (or deceleration) can increase the likelihood or the severity of the natural risks to the system (collapse, disintegration, etc.). This method involves techniques that can detect and offset or neutralize such risks, both before and if they materialize. So called "leading indicators" are important in this regard. These are acknowledged statistical precursors of more serious situations in complex-systems. Another way to think of this method is policing its autonomous agents.

But all of the methods are still generalizations. For most engineers, until there are explicit examples, these methods will remain cryptic. Unfortunately, brevity precludes a discussion of any example of any of these methods in any sort of detail.

As an illustration, just one of the ways that an engineer can formulate and apply stimulants that promote synergistic self-directed development in a complex-system that is also a social system is to facilitate the appearance or spread of markets. There are various sorts of markets: for example brokered markets and unregulated markets. Each has its strengths and liabilities. Given a particular complex-system, a brokered market may seem to be the most advantageous or practical. So, how does one go about facilitating a brokered market in such a case? And given that such a market develops, what does one do then? Here are some of the issues. It is necessary to identify autonomous agents that might function as buyers and sellers – and of course as a broker. It might even be necessary to create or inject an agent that will assume the role of broker. It will be necessary to characterize the goods or services that might be exchanged. It will be necessary to characterize the value propositions that will be used to mediate exchanges in the market. And it will be necessary to identify the impediments to such exchanges and how a broker can neutralize such impediments at a profit to itself without undermining the propensity for exchanges between buyers and sellers. Brokered markets succeed because they are win-win-win propositions. Complexity is being accumulated. Needless to say, answering these and other questions in any detail is context sensitive. This is well beyond the scope of this introduction.

PULLING IT TOGETHER (PART 2)

When TSE is not applicable, cSE will be. Since the system in question will be one that is evolving, the very first thing to be done is to characterize the system as a set of populations at multiple scales of conceptualization, and then how the predicates of evolution are being expressed in those populations. This includes the five classical predicates; but it should also include attention to the less familiar ones such as characterizing the life cycles of the members of critical populations. Beyond this, engineering estimates for the complexity and fitness of the populations and their members should be developed and maintained as the populations evolve. And Price's equation should be used in relating fitness to the properties and relations of the members of populations.

Multi scale analysis is applied as well – but not just once. This is done repeatedly as the system continues to evolve. In doing this it is important to be clear about characterizing the relevant scales of conceptualization involved, including as needed the associated frames of reference. Second, reductionist techniques can be applied at each individual scale of conceptualization. This includes analysis based on Sarnoff and Metcalf network models. But this might not always be feasible even if it is theoretically possible. Applying reductionism depends on completeness and detailed thoroughness. The effort required can be computationally prohibitive. And even if it is practical to apply reductionist methods, the significance may not be great. It is the autonomous agents in a complex-system that, if present, drive its development – and reductionism seldom provides useful insights into their motivations. Third, it is also possible to analyze the single scale aspects of systems using regime analysis. Radial Category Theory and Reed networks will aid in this regard, as will straightforward statistical analysis. Fourth, in terms of cross scale analysis, the only available techniques today are those of statistical analysis. This will reveal correlations and dependencies – but generally not causality. All of this will be important to the extent that the analysis informs the application of the Regimen of cSE.

And, of course, the appropriate forms of the methods in the Regimen must be formulated and applied as the system evolves.

Finally, it is always important to keep in mind that engineering a complex-system precludes direct control. It is the system itself that is in control, not the system engineer. The system engineer will be accountable, but he or she will not be in control. If it is initially difficult to understand what this might mean, then the example of a teacher and his or her student should be considered. It is the student who must learn. The teacher can only teach. But both will be held accountable.

CONCLUSION

Now, if it is desired to know more about cSE, or if there is an interest in applying what is already known, then call me. I might be able to help. If this is not appropriate, there are alternatives. It is possible, for example, to figure out how to do cSE on your own. A reading list (bibliography) is included to use in getting started. And, of course, this introduction is available as well.

BIBLIOGRAPHY

Anderson, P. W. (1972, August 4). More is Different. *Science, New Series vol. 177 no. 4047* .

Gladwell, M. (2000). *The Tipping Point.* Little, Brown, and Company.

Heylighen, F., Cilliers, P., & Gershenson, C. (April, 2006). *Complexity and Philosophy.* arXiv.org pre-print.

Kuras, M. L. (2004). *MP 04B000038: Complex-system Engineering.* Bedford, MA: The MITRE Corporation.

Kuras, M. L. (2006). *MTR 06B000060: A Multi Scale Definition of a System.* Bedford, MA: The MITRE Corp.

Lakoff, G. (1987). *Fire, Women, and Dangerous Things.* University of Chicago Press.

Malone, T. W. (2004). *The Future of Work.* Harvard Business School Press.

Marwell, G., & Oliver, P. (1993). *The Critical Mass in Collective Action*. Cambridge University Press.

Okasha, S. (2006). *Evolution and the Levels of Selection*. Oxford: Clarendon Press.

Surowiecki, J. (2004). *The Wisdom of Crowds.* Anchor Books.

Weinberg, G. M. (1975; 2001). *An Introduction to General Systems Thinking.* Dorset House Publishing.

Chapter 18

Statistical properties of agent-based market area model

Zoltán Kuscsik, Denis Horváth

Department of Theoretical Physics and Astrophysics, University of
P.J. Šafárik, Košice, Slovak Republic

One dimensional stylized model taking into account spatial activity of firms with uniformly distributed customers is proposed. The spatial selling area of each firm is defined by a short interval cut out from selling space (large interval). In this representation, the firm size is directly associated with the size of its selling interval.

The recursive synchronous dynamics of economic evolution is discussed where the growth rate is proportional to the firm size incremented by the term including the overlap of the selling area with areas of competing firms. Other words, the overlap of selling areas inherently generate a negative feedback originated from the pattern of demand. Numerical simulations focused on the obtaining of the firm size distributions uncovered that the range of free parameters where the Pareto's law holds corresponds to the range for which the pair correlation between the nearest neighbor firms attains its minimum.

1 Introduction

The study of elemental interactions in social and economical systems has a great importance to understand the large-scale system properties. One of the universal large-scale properties exhibited by social systems in a robust way is the *Pareto's law* of wealth distribution and firm size [3, 6, 10]. Pareto's law is generally associated to the observation, that personal income of individuals, the size of companies are distributed by power-law.

The formation of power-laws has generally complex origin. Among other approaches, highly sophisticated multi-agent models have been developed [2, 14, 15] to explain the power-laws observed in various social systems.

In this paper we propose agent-based model that emphasizes role of spatial location of firm within the limited market area. The model approximates the basic mechanisms of competency that simply follows from the spatial positions and selling activities of firms.

An extensive economic literature exists that deals with the competitiveness as consequence of location. One dimensional model of spatial duopoly introduced by Hotelling [8] has assumed that consumers are continuously and uniformly distributed along a line segment. The model of firm distribution in a non-uniform environment has been developed by Lawrence [16]. It predicts firm density in an urban setting in which the population density decreases exponentially with the distance from the center. Erlenkotter [5] has considered uniformly distributed demand over the infinite plane. He has discussed various regular two-dimensional market shapes. An elegant and advanced multistore competitive model of two firms in a finite business area has been introduced by Dasci and Laporte [4]. This model has been investigated for one and two dimensional geographical markets. It assumes the costumers are dispersed through space in only one direction along some coordinate $x \in (0, 1)$. In this regard it is useful to mention the functional expression for total revenue per firm

$$\int_0^1 Q(x)f(x)\mathrm{dx}, \tag{1}$$

where $f(x)$ is the probability density function of the customers multiplied by the probability $Q(x)$ that customer at x patronizes product of given firm.

As we have mentioned before, our present approach also pays attention to spatial aspects. The approach comes from ecologic-economic feedback concept of regulated factory emissions introduced by us recently [9]. The work points out an emergence of critical properties in a two dimensional system with spatially distributed agents balancing the conflicting objectives. The model assumes that sources of diffusive emissions compete with the distributed sensorial agents. The analysis of the complex numerical data yield us to reductionist and purely geometric formulation that is related to coverage percolation problem. The geometric idea has been applied here to study the spatial distribution of the competitive firms reduced to basic geometric objects that cover market area. Our stylized spatial model is defined as it follows.

2 Firm growth

We assume that each firm acts as a seller agent of a product from the same sort of industry or it behaves as a provider of some service business. The spatial economic activity of the ith agent is defined by its position $x_i^{(t)} \in (0, L)$ and by its selling area $(x_i^{(t)} - r_i^{(t)}, x_i^{(t)} + r_i^{(t)})_{\mathrm{mod}L}$, where L is the constant size of one dimensional market space with periodic boundary conditions and $r_i^{(t)}$ is the selling radius of firm i. It should be noted that radius does not mean strictly the space of the physical activity of the seller but it can be understood as a radius up to which the customers are attracted. We have considered customers uniformly distributed along a straight line. Interestingly, such arrangement is typical for the restaurants distributed along a main road or highway [4].

We follow with definition of the measure of the spatial activity of the ith firm

$$s_i^{(t)}(x) = \begin{cases} 1 & \text{if } i \text{ claims to sell at the position } x \\ 0 & \text{if } x \text{ is not from the agent's selling area} \end{cases} \tag{2}$$

Generally, large selling area means more potential consumers covered by delivery of given product which results higher interest connected to higher profit. Without negative economic feedback, the continuous investment of the constant fraction of income yields to the exponential growth of the firm size and its sale. Several facts that yield the negative feedback between firm and its environment should be mentioned: (i) the transportation costs of products are convex functions of distance [11]; (ii) the complexity of firm management grows with a firm size (iii) larger firms use more sophisticated and thus more expensive information technologies; (iv) the presence of two or more competitive products in the same location affects the prices as well as the annual sales.

Here we assume only a negative feedback that originates from the spatial *overlap* of selling areas. The overlap of ith firm area with the areas of the remaining $(N-1)$ firms is defined by

$$\Omega_i^{(t)} = \sum_{\substack{j=0 \\ j\neq i}}^{N} \int_0^L s_i^{(t)}(x)s_j^{(t)}(x)\mathrm{d}x. \tag{3}$$

With this firm-firm interaction picture in mind, we suggested the dynamical rule of the firm growth

$$r_i^{(t+1)} = r_i^{(t)} + \alpha r_i^{(t)} - \beta \Omega_i^{(t)}, \tag{4}$$

where $\alpha > 0$ and $\beta > 0$ are constant parameters controlling the instantaneous growth. The term $-\beta \Omega_i^{(t)}$ can be interpreted as a negative feedback that reflects the competition. The *selling area* of firm i is expressed by

$$S_i^{(t)} = \int_0^L s_i^{(t)}(x)\,\mathrm{d}x = 2r_i^{(t)}. \tag{5}$$

3　Firm establishment and bankruptcy

In the stylized version of the model we study the firm is established at random position with a small random initial selling radius $r_i^{(t)} \in (r_a, r_b)$, where r_a is assumed to be the lower bound of profitability (smallest firm). Therefore, the bankruptcy of a firm occurs when $r_i^{(t)} < r_a$. At the same time new firm (with the same index) is established at a new random position with some initially random size. This death-birth process is analogous to the so called extremal dynamics principle [1] applied to e.g. models of the wealth distribution [12] and stock markets [7].

4 Simulation and Numerical results

The choice of parameters is chronic problem specially in models of the social-economic systems. Although models of interacting agents give qualitative predictions that in many aspects resemble behavior of real-world systems, in the most cases, the quantitative analysis needs laborious tuning of parameters until the range is reached for which the phenomenon of interest takes place. Our simulations were performed with constant number of firms $N = 500$ for predefined market area $L = 3 \times 10^5$. The constant growth factor $\alpha = 0.01$ and the initial range of the selling space constrained by radii $r_a = 2.0$ and $r_b = 5.0$ is chosen to invoke steps much smaller than L.

To reach the stationary regime 10^5 the initial synchronous system updates were discarded. The information from subsequent 10^6 updates spent in stationary regime has been recorded. Their analysis has uncovered that firm-firm interaction controlled by β admits to establish market regimes that differ in size distributions. We observed that sufficiently large β leads to the market with lowered overlaps. On other hand, sufficiently small β supports formation of oligopoly that cover a dominant area of available market space. The important for us power-law distributions are observable only for exceptional β (see fig 4). This finding opens a question: what regulatory real-world economic principle controls the sustaining of the empirically relevant power-law regimes. The related question is the optimization of free parameters. For this purpose we examined several heuristic criteria. The most attractive seems to be an extremal entropy principle [13], but in that case one faces to the usual problem of the proper entropy definition.

More pragmatic, however, less fundamental attempt comes from our analysis of firm-firm correlations. For this purpose the pair correlation function C of sizes of nearest neighboring firms (k) at positions $x_i^{(t)} \le x_k^{(t)}$ can be defined as

$$C = \left\langle \frac{\frac{1}{N}\sum_{i=1}^N r_i^{(t)} r_k^{(t)} - \left(\frac{1}{N}\sum_{i=1}^N r_i^{(t)}\right)^2}{\frac{1}{N}\sum_{i=1}^N (r_i^{(t)})^2 - \left(\frac{1}{N}\sum_{i=1}^N r_i^{(t)}\right)^2} \right\rangle_t. \tag{6}$$

Here $\langle \ldots \rangle_t$ denotes temporal stationary average. The calculations for different β uncovered that minimum of correlation function corresponds to parameter (or narrow interval of parameters) for which nearly power-law distributions can be fitted quite well. This extremal principle reflects the possible importance of the measurements of the spatial correlations in social and economic systems.

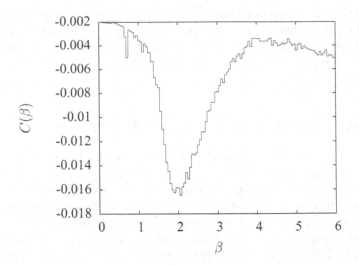

Figure 1: Plot of the pair correlation function defined by Eq. 6 as a function of β. The Pareto's law for the firm size belongs to β where $C(\beta)$ attains its minimum.

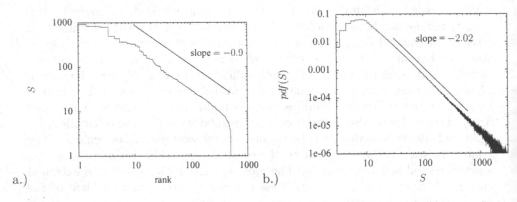

a.)   b.)

Figure 2: a.) Plot of sizes of firms as a function their rank for $\beta = 2.0$. The fitted power-law index is close to unity. b.) The distribution of firm sizes.

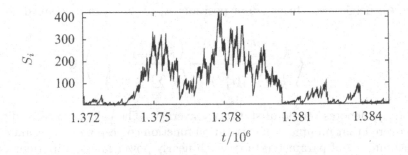

Figure 3: The time evolution of size of selected firm.

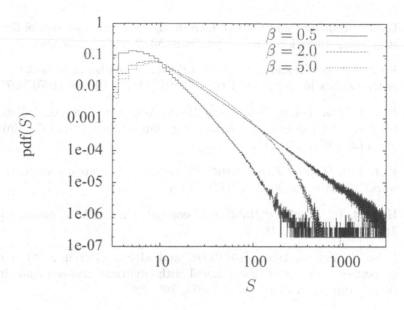

Figure 4: Distribution of firm sizes plotted for different β.

5 Conclusions

By focusing on the geometric representation of firms we proposed a stylized multi-agent model of firm growth. The competitive dynamics of firms under which the system reaches a steady state results a complex patterns of firm locations. Despite of its formal simplicity, the model supplemented by the principle of minimum firm-firm correlation is able to explain the origin of the Pareto's law. Further validating of model is planned that takes into account real-world data. Hoverer, this will need to take into account the non-uniform distribution of customers. The advanced model of this type is under development.

Acknowledgments The authors would like to express their thanks to Slovak Grant agency LPP APVV (grant no. 0098-06), VEGA (grant no. 1/4021/07, 1/2009/05) for financial support.

Bibliography

[1] BOETTCHER, Stefan, and Allon G. PERCUS, "Optimization with extremal dynamics", *Phys. Rev. Lett.* **86**, 23 (Jun 2001), 5211–5214.

[2] BOUCHAUD, Jean-Philippe, and Marc MEZARD, "Wealth condensation in a simple model of economy", *Physica A* **282** (Jul 2000), 536–545.

[3] CLEMENTI, F., T. DI MATTEO, and M. GALLEGATI, "The power-law tail exponent of income distributions", *Physica A* **370**, 1 (Oct 2006), 49–53.

[4] DASCI, Abdullah, and Gilbert LAPORTE, "A continuous model for multi-store competitive location", *Oper. Res.* **53**, 2 (2005), 263–280.

[5] ERLENKOTTER, Donald, "The general optimal market area model", *Annals of Operations Research* **18**, 1 (Dec 1989), 43–70, 10.1007/BF02097795.

[6] HEGYI, Geza, Zoltan NEDA, and Maria AUGUSTA SANTOS, "Wealth distribution and pareto's law in the hungarian medieval society", *Physica A* **380** (Jul 2007), 271–277.

[7] HORVATH, D., and Z. KUSCSIK, "Structurally dynamic spin market networks", *ArXiv - physics/0701156* (2007).

[8] HOTELLING, Harold, "Stability in competition", *The Economic Journal* **39**, 153 (Mar 1929), 41–57.

[9] KUSCSIK, Zoltan, Denis HORVATH, and Martin GMITRA, "The critical properties of the agent-based model with environmental-economic interactions", *Physica A* **379**, 1 (Jun 2007), 199–206.

[10] MALEVERGNE, Yannick, and Didier SORNETTE, "A two-factor asset pricing model and the fat tail distribution of firm sizes", *SSRN eLibrary* (2007).

[11] MCCANN, Philip, "Transport costs and new economic geography", *J Econ Geogr* **5**, 3 (2005), 305–318.

[12] PIANEGONDA, S., J. R. IGLESIAS, G. ABRAMSON, and J. L. VEGA, "Wealth redistribution with conservative exchanges", *Physica A: Statistical Mechanics and its Applications* **322** (May 2003), 667–675.

[13] RAINE, Alan, John FOSTER, and Jason POTTS, "The new entropy law and the economic process", *Ecological Complexity* **3**, 4 (Dec 2006), 354–360.

[14] RAWLINGS, Philip K., David REGUERA, and Howard REISS, "Entropic basis of the pareto law", *Physica A* **343** (Nov 2004), 643–652.

[15] SAMANIDOU, E, E ZSCHISCHANG, D STAUFFER, and T LUX, "Agent-based models of financial markets", *Reports on Progress in Physics* **70**, 3 (2007), 409–450.

[16] WHITE, Lawrence J., "The spatial distribution of retail firms in an urban setting", *Regional Science and Urban Economics* **5**, 3 (August 1975), 325–333.

Chapter 19

Complexity and the Social Sciences: Insights from complementary theoretical perspectives

A. J. Masys

University of Leicester, UK

The application of complexity theory crosses many domains thereby reflecting the multidisciplinary perspective inherent within the concept. Within the social sciences, the advent of complexity theory has facilitated a re-examination of the concept of system, '...rejecting old assumptions about equilibrium in favour of the analysis of dynamic processes of systems far from equilibrium, and respecifying the relationship of a system to its environment' (Walby, 2003).

The term 'System accidents' describes an aetiology that arises from the interactions among components (electromechanical, digital, and human) rather than the failure of individual components. Accidents involving complex socio-technical systems, such as that resident within the nuclear power industry, aerospace industry and military operations, reflect this aetiology characterized by its nonlinearity and inherent complexity. "Complex systems cannot be understood by studying parts in isolation. The very essence of the system lies in the interaction between parts and the overall behaviour that emerges from the interactions" (Ottino, 2003). The application of Actor Network Theory (ANT) facilitates an examination of complex socio-technical systems focusing on the interconnectedness of the heterogeneous elements characterized by the technological and non-technological (human, social, organizational) elements that comprise the problem space. This paper presents and argues for the integration of complexity theory as a complementary theoretical perspective to the field of sociology as a means of generating insights and increasing explanatory and conceptual depth of analysis. The integration of ANT and complexity theory in analyzing aviation accident aetiology is presented as an example.

1   Introduction

The victims and fallout associated with the tragic events of Bhopal, Three Mile Island, Chernobyl, Challenger Space Shuttle and more recently the Columbia Space Shuttle are constant reminders of the dangers associated with complex socio-technical systems.

With the advent of complex coupled systems and the evolutionary introduction of new technology, the aetiology of accidents is changing. 'Since World War II, we are increasingly experiencing a new type of accident that arises in the interactions among components (electromechanical, digital, and human) rather than the failure of individual components. Perrow (1984) coined the term 'system accident' to describe it' (Leveson, 2003).

These failures manifest not as a result of some simple unforeseen cause, but 'from highly complex human activity systems containing large numbers of interconnected subsystems and components' (Midgley, 2003:320). Bennett (2002:1) describes the aetiology of accidents involving complex socio-technical systems as:

> 'Failure, whether human, technological or corporate, is a complex phenomenon. It often arises out of unforeseeable interactions between system components or systems. The seeds of failure may have been sown years-decades , even before malfunction or collapse. Failure may originate in a complex interplay between social, economic and political factors.'

The technical perspective of accident aetiology is rooted within the probability of failure models associated with components of a system. This perspective traces the failure of a system to a chain of events within a system that linearly define the path towards an accident.

The systems perspective challenges the hegemony of traditional accident causation models characterized by a paradigm of a 'chain of failures'. According to Leveson (2002) '...viewing accidents as chains of events may limit understanding and learning from the loss. Event chains developed to explain an accident usually concentrate on the proximate events immediately preceding the loss. But the foundation for an accident is often laid years before.'

The body of knowledge within the social sciences regarding accident aetiology of complex socio-technical systems has increased over the last 20 years recognizing the social and technical dimensions of accidents. The anatomy of disasters and accidents involving socio-technical systems show an aetiology that reflects an inherent complexity that involves elements beyond the temporally and spatially proximate thereby supporting a systemic view of disasters and accidents. The reductionist paradigm that focused on the parts of a system and how they functioned is replaced by a paradigm that embraces the complex. We now focus on the interrelationships and the interactions of the actors in an analysis of the behaviour and topology of the system.

From a methodological standpoint, we must move beyond the view of the system as simply 'a whole equal to the sum of its parts' and consider the interrelations and causal effects, which are often complex and nonlinear thereby shedding light on the 'system effects' such as emergence, equifinality and mutlifinality. Altmann and Koch (1998:183) remark that 'if the system is analyzed only in terms of its parts, as assumed in atomism and more generally in reductionism, the system-effects are lost without trace'.

Sociology offers an interesting approach to looking at the socio-technical elements of complex systems through the application of Actor Network Theory (ANT). The systems perspective of ANT looks at the inter-connectedness of the heterogeneous elements characterized by the technological and non-technological (human, social, organizational) elements. The network space of the actor network provides the domain of analysis that presents the accident aetiology resident in a network of heterogeneous elements that shape and are shaped by the network space.

Germane to this work, the socio-technical system is a topic of inquiry within sociology that combines the social and technical paradigms and examines the relationship between them. As described by Coakes (2003:2), 'Socio-technical thinking is holistic in its essence; it is not the dichotomy implied by the name; it is an intertwining of human, organizational, technical and other facets'. Senge argues that since the world exhibits qualities of wholeness, the relevance of systemic thinking is captured within its paradigm of interdependency, complexity and wholeness (Flood, 1999). Although events can be considered to be discrete occurrences in time and space '...they are all interconnected. Events, then, can be understood only by contemplating the whole' (Flood, 1999:13).

Through analysis of various aviation accidents and in particular cases of fratricide (air-toground), the application of the ANT perspective and concepts from complexity theory revealed characteristics of the 'problem space' that helps us understand accident aetiology. The inherent complexity revealed through this perspective focuses on such characteristics as interrelations, interconnectivity and the dynamic nature of the relations that shape the 'social'. The complementary application of complexity theory as a framework helps us to understand the complexity inherent within the 'social'. Complexity theory provides a new set of conceptual tools to help address the classic dilemmas of social science, facilitating new ways of thinking of 'system' as well as challenging the reductionist perspective so resident in scientific enquiry (Walby, 2003). This is particularly germane to our sociological analysis of accident aetiology involving complex socio-technical systems.

2 Complexity Thinking

Complexity theory is an interdisciplinary field of research that has become recognized as a new field of inquiry focusing on understanding the complexity inherent within the behaviour and nature of systems.

As applied to the social sciences, complexity theory, complexity thinking provides a perspective of the 'social world' that reveals emergent properties, nonlinearity, consideration of the 'dynamic system', interactions, interrelations that is transforming the traditional views of the social as reflected in '...Guastello (1995), Dooley (1997), Eoyang (1997), McKelvey (1997), Zimmerman et al. (1998), Anderson (1999), Poole et al. (2000)' (Dooley et al, 2003).

Addressing issues that lie at the foundation of sociological theory, complexity theory facilitates '...a re-conceptualization and re-thinking regarding the nature of systems reflecting dynamic inter-relationships between phenomenon. The new theorizations of system within complexity theory radically transform the concept making it applicable to the most dynamic and uneven of changing phenomena' (Walby, 2003:3).

Important features that characterize complex systems and their behaviour include the ability to produce properties at the collective level that are not present when the components are considered individually as well as their sensitivity to small perturbations. This dynamic behaviour of complex systems involves interactions at all scales. The complexity inherent within the system may result in changes in behaviour or topology that only become discernable at the macroscale, thereby making the analysis of accident aetiology problematic.

The features of systems thinking and complexity theory that shape the methodological approach associated with this work stem from the conceptualization that the general system is not simply an aggregation of objects but rather is a set of interrelated, interconnecting parts creating through their interaction new system properties, which do not exist in a loose collection of individual objects. The realization of this reification corrects many of the failures and mistakes of classical and modern science, 'especially in their attempts to understand and explain complex phenomena and processes' (Altmann and Koch, 1998:186).

Germane to this work is the departure from linear models of accident causation to the acknowledgement of an inherent non-linearity of complex socio-technical systems thereby recognizing the multiplicities whose causes and effects are always dependent on a variety of influences.

Complexity theory thereby provides a framework that facilitates sociological analysis focusing on systems recognizing qualities of nonlinear dynamic behaviour and emergence.

3   Actor Network Theory

As discussed in detail in Masys (2004, 2005), 'ANT . . . is a relatively new, and still rapidly developing, direction in social theory that has emerged from post structuralism the writings of Foucault and Deleuze (and Guattari) in particular and sociological studies of science (e.g. laboratory studies) and technology with the writings of Serres, Latour and others such as Callon and Law being particularly significant (Smith, 2003)'.

'ANT was developed to analyse situations in which it is difficult to separate human and non-humans, and in which the actors have variable forms and competencies' (Callon 1999:183).

Focusing on the socio-technical domain, ANT views the world as heterogeneous and thereby rejects the 'artificial' schism between the social and the technical as illustrated in Latour (1993), Callon (1986a, 1986b), Law (1987), Law and Callon (1988).

Latour (2005) refers to the 'social' as a trail of associations between heterogeneous elements '...a type of connection between things that are not themselves social'.

Recasting the social in terms of associations, relations, characterized by an inherent heterogeneity, and complex interconnectivity is a fundamental paradigm shift that challenges the traditional understanding of 'social' and in so doing facilitates insight into the world of complex socio-technical systems.

Fundamental concepts of ANT are the conceptualization of the Actor and the Network.

An actor-network, as cited in Aanestad and Hanseth (2000), 'is a heterogeneous network of human and nonhuman actors... where the relations between them are important, rather than their essential or inherent features (Latour, 1987; Callon, 1986, 1991).'

The actor, whether technical or non-technical, is examined within the context of a heterogeneous network (Aanestad and Hanseth, 2000).

The choice to use ANT as a theoretical framework for the analysis stems from its ability to analyze occasions offered by accidents and breakdowns whereby as Latour (2005:81) remarks '...all of a sudden, completely silent intermediaries become full-blown mediators; even objects, which a minute before appeared fully automatic, autonomous, and devoid of human agents, are now made of crowds of frantically moving humans with heavy equipment. Those who watched the Columbia shuttle instantly transformed form the most complicated human instrument ever assembled to a rain of debris falling over Texas will realize how quickly objects flip-flop their existence. Fortunately for ANT, the recent proliferation of 'risky' objects has multiplied the occasions to hear, see, and feel what objects may be doing when they break other actors down.' ANT considers both social and technical determinism as flawed and thereby suggests a position that takes into consideration the socio-technical perspective such that neither the social nor technical are privileged. Hence ANT, as a theoretical perspective for this work provides a mechanism to examine accident aetiology from a systems viewpoint embracing a complexity paradigm. The properties associated with the ANT perspective allow one to approach the accident analysis without privileging either the social or technical elements recognizing the interconnectivity, nonlinearity and emergent behaviour that so characterizes accident aetiology and resides within complexity thinking.

4 Discussion

The ANT perspective that views the 'social' as an emergent characteristic of the network space complemented by the framework of complexity theory challenges the traditional cause and effect paradigm that is resident within the technical based approaches to accident aetiology. As an integrating element, complexity theory provided not a methodology per se, but rather 'a conceptual framework, a way of thinking, and a way of seeing the world' (Mitleton-Kelly, 2004).

A core feature of complexity theory is its fundamental re-thinking of the nature of systems, recognizing the simultaneously dynamic and systemic inter-relationships and interconnectivity. Dekker (2005:31) points to the requirement for a systems perspective with regards to understanding accident aetiology. He asserts that:

> 'Systems' thinking is about relationships and integration. It sees a socio-technical system not as a structure consisting of constituent departments, blunt ends and sharp ends, deficiencies and flaws, but as a complex web of dynamic, evolving relationships and transactions. ...Understanding the whole is quite different from understanding an assembly of separate components. Instead of mechanical linkages between components (with a cause and an effect), it sees transactionssimultaneous and mutually interdependent interactions. Such emergent properties are destroyed when the system is dissected and studied as a bunch of isolated components (a manager, department, regulator, manufacturer, operator). Emergent properties do not exist at lower levels; they cannot even be described meaningfully with languages appropriate for these lower levels'.

Through the application of Complexity Theory we are introduced to non-linear processes such that small changes in inputs can have dramatic and unexpected effects on outputs. As articulated by Urry (2004):

> 'This complex systems world is, according to Axelrod and Cohen, a world of avalanches, of founder effects, or self-restoring patterns , of apparently stable regimes that suddenly collapse, of punctuated equilibria, of 'butterfly effects', and of thresholds as systems suddenly tip from one state to another'.

It is from this analysis (integrating ANT and Complexity Theory) that the hegemony of 'pilot error' is dispelled and replaced by an aetiology characterized by a 'de-centered causality'. The inherent complexity revealed through this perspective focuses on such characteristics as interrelations, interconnectivity and the dynamic nature of the relations that shape the 'social' as an emergent property. The actors within this network space do not preexist, rather they emerge as a result of an entangled interconnectivity, or as Barad (2007) argues as an 'intra-relating'.

The analysis methodology of ANT, 'Following the actors', revealed the notion of a complex co-evolving ecosystem characterized by '...intricate and multiple intertwined interactions and relationships. Connectivity and interdependence propagates the effects of actions, decisions and behaviours ..., but that propagation or influence is not uniform as it depends on the degree of connectedness' (Mitleton-Kelly and Papaefthimiou, 2000).

Revealed in the analysis and as discussed in detail in Masys (2006), Foucault's notion of disciplinary power helps to explain the inscription and translation processes within ANT as applied to socio-technical systems. The analysis of the inherent interconnectivity within complex socio-technical systems reaffirms Foucault's perspective that power is embodied in heterogeneous micro-practices and power is seen as enacted and discontinuous rather than stable and exercised by a central actor (Thompson and McHugh, 1995) (cited in Rolland and Aanestad, 2003). As articulated by Yeung (2002),

> 'Actors in these relational geometries are not static "things" fixed in time and space, but rather agencies whose relational practices unleash power inscribed in relational geometries and whose identities, subjectivities, and experiences are always (re)constituted by such practices.'

Further as articulated by Yeung (2002),

> '...causal power can be ascribed to relational networks when their relational geometries generate an emergent effect such that the sum of these relations is much greater than that of individual actors. The geometrical configurations of these emergent network relations provide the central dynamic to drive networks and to produce spatial outcomes. Power is thus constituted collectively by network relations and its influence can only be realized in a relational sense through the exercise of its capacity to influence. Actors in relational geometries do not possess power per se. Through their practice, actors perform the role as the agents exercising that emergent power inscribed in relational geometries.'

Combining ANT and complexity thinking with Foucault's conceptualization of power, highlights how the micro-practices 'constantly get configured and reconfigured as 'disciplinary technologies' (Aanestad and Hanseth, 2000), as reflected in design and organizational decisions. The 'hardwired politics' and power emerge as the 'deleted voices' that permeate the network space, the relations and the actors and shape decisionmaking and the accident aetiology.

The network space 'worldview' captures the system perspective of aviation accident aetiology revealing the 'social' in terms of this 'hardwired politics'. The analysis revealed that the term pilot error is pejorative, a reflection of an event-based approach to accident causation that fails to capture the nonlinearity inherent within accident aetiology. A de-centered causality dispels the hegemony of 'pilot error', thereby facilitating an accident 'model' that is characterized as dynamic, nonlinear with emergent properties and embracing features resident within complexity theory. This entails a rethinking of some fundamental concepts such as notions of causality, agency, power, space and time. As Barad (2007: 394) argues, '...future moments don't follow present ones like beads on a string. Effect does not follow cause hand over fist...causality is an entangled affair.' Deleuze's perspective of space-time as "folded, crumpled, and multi-dimensional" (Deleuze,1995) is germane to the conceptualization of ANT. It is this very nature of this space-time schema that "the coordinates of distance and proximity are transformed by a folding, refolding, and unfolding that eschews ideas such as linearity" (Smith, 2003).

As Latour (1996:238) remarks '...the social as actor network is hybrid: it is a heterogenesis that consists of discursive, human, and material elements, which simultaneously coexist, and which cannot be separated from one another'. Complexity theory, as a complementary theoretical perspective to sociology, increases the conceptual depth of analysis and understanding regarding accident aetiology involving complex socio-technical systems.

5   Conclusion

The application of systems theory facilitates a foundational perspective that guides the development of an accident aetiology model based on insights from ANT and complexity theory. Through an analysis facilitated by ANT, the hegemony of 'blamism' associated with 'pilot error' is replaced with a de-centered accident aetiology that is revealed within a network of heterogeneous elements. Hard-wired politics (reified as 'illusions of certainty') and power characterize the network space; thereby shaping the aetiology of the accidents and transcending the linearity associated with traditional understanding of accident aetiology. The contribution of complexity thinking to sociology and in particular to the study of complex socio-technical systems is reflected in our attention to the dynamic processes, interconnectivity and relationality facilitating a rethinking regarding the concept of system. The application of complexity theory facilitates a paradigm shift complementing the perspective of ANT and systems theory. Together ANT and Complexity Theory provides a framework for thinking through larger implications of theories detailing accident aetiology. It reworks our understanding of what happens in accidents. The results of this work are far reaching in terms of how we view sociotechnical systems and accident aetiology.

6   References

[1] Aanestad, M. and Hanseth, O. (2000). "Implementing open network technologies in complex work practices: a case from telemedicine", in Baskerville, R.E., Stage, J. and DeGross, J.I. (Eds), Organizational and Social Perspectives on information Technologies, Kluwer Academic Publishers, Dordrecht, 2000, pp. 355-69.

[2] Altmann, G., Koch, W.A. (1998) Systems: New paradigms for human sciences. Gruyter, New York,1998. pg 183.

[3] Barad, K. (2007) Meeting the Universe Halfway:quantum physics and the entanglement of matter and meaning. Duke University Press. Durham & London 2007.

[4] Bennett, S. (2002)Victims of Hubris? The Decline and Fall of Pan Am and Swissair. Scarman Centre, Leicester.

[5] Callon, M. (1986a). The Sociology of an Actor-Network: The Case of the Electric Vehicle. Mapping the Dynamics of Science and Technology. Calon, M., Law, J. and Rip, A. (Eds). Macmillan Press, London: 19-34.

[6] Callon, M. (1986b). Some of the Elements of a Sociology of Translation: Domestication of the Scallops and the Fishermen of St. Brieuc Bay. Power, Action & belief. A New Sociology of Knowledge? Law, J. (Ed). Routledge & Kegan Paul, London: 196-229.

[7] Coakes, E. (2003) Socio-technical Thinking- An Holistic Viewpoint. In Clarke, S., Coakes, E., Hunter, M.G., Wenn, A. Socio-technical and human cognition elements of information systems. Information Science Publishing. Hershey.pg 2

[8] Dekker, S. (2005) Ten Questions about human error. A new view of human factors and system safety. Lawrence Erlbaum Associates, Inc. New jersey. pg 10.

[9] Dooley, K.J., Corman, S.R., McPhee, R.D., Kuhn, T. (2003) Modeling High-Resolution Broadbrand Discourse in Complex Adaptive Systems. Published in Nonlinear Dynamics, Psychology, & Life sciences. Vol. 7, number 1, January 2003. pp. 61-85.

[10] Flood, R.L. (1999) Rethinking the fifth discipline: Learning within the unknowable. Routledge, London. pg 13

[11] Latour, B. (1993) We have never been modern. Harvester Wheatsheaf, Hemel Hempstead.

[12] Latour,B. (2005) Reassembling the Social: An Introduction to Actor network theory. Oxford University Press.

[13] Law, J. (1987). Technology and Heterogeneous Engineering: The Case of Portuguese Expansion. The Social Construction of Technological Systems: New Directions on the Sociology and Hostory of Technology. Bijker, W.E., Hughes, T.P. and Pinch, T.J. (Eds). MIT Press, Cambridge, Ma: 111-134.

[14] Law, J. and Callon, M. (1988). Engineering and Sociology in a Military Aircraft Project: A Network Analysis of technological Change. Social Problems 35(3): 284-297.

[15] Leveson, N.G. (2002) System Safety Engineering: Back to the Future. Aeronautics and Astronautics, MIT.

[16] Leveson, N. (2004) A New Accident Model for Engineering Safer Systems. Safety Science, Vol. 42, No. 4, April, pp. 237-270

[17] Masys, A.J. (2004) Aviation Accident Aetiology: A systemic examination of Fratricide using Actor Network Theory. Proceedings of the 22nd International System Safety Conference (ISSC), Providence, RI, 2-6 August 2004. pp. 413-423.

[18] Masys, A.J. (2005) Aviation Accident Aetiology: A Systemic Examination of Fratricide using Actor Network Theory. Published in the Proceedings of the International System Safety Conference, Providence, Rhode Island. 2005.

[19] Masys, A.J. (2006) Understanding Fratricide. Insights from Actor Network Theory and Complexity Theory. Published in the Proceedings of the International System Safety Conference, 2006. Albequere, New Mexico. 31 July-5

August 2006.

[20] Midgley, G. (2003) Systems Thinking. Sage Publications. London.

[21] Mitleton-Kelly,E., Papaefthimiou, M.C.(2000) Co-evolution of diverse elements interacting within a social ecosystem. Feast 2000 International Workshop on 'Feedback and Evolution in Software and Business Processes', Imperial College, London, UK, July 10-12, 2000.

[22] Mitleton-Kelly, E. (2004) Complex systems and evolutionary perspectives on Organisations: The Application of Complexity Theory to Organisations. `http://www.psych.lse.ac.uk/complexity/PDFiles/publication/EMK_The_Principles_of_Complexity.pdf`

[23] Ottino, J. (2003), "Complex systems", AIChE Journal, Vol. 49 No. 2.

[24] Rolland, K.H., Aanestad, M. (2003) The Techno-Political Dynamics of Information Infrastructure Development: Interpreting Two Cases of Puzzling evidence. Retrieved from `http://heim.ifi.uio.no/~margunn/2003/IRIS26.pdf`

[25] Smith, R.G. (2003) World City Actor-Networks. Progress in Human Geography, 27 (1) 2003.

[26] Urry, J. (2004) Small Worlds and the new social physics. Global networks 2004.pg 3

[27] Walby, S. (2003) Complexity Theory, Globalisation and Diversity. Presented to the conference of the British Sociological Association, University of York, April 2003.

[28] Yeung, H.WC. (2002) Towards a Relational Economic Geography: Old Wine in New Bottles? Paper Presented at the 98th Annual meeting of the Association of American Geographers, Los Angeles, USA, 1923 March 2002.

Chapter 20

Agent Based Model of Livestock Movements

D.J. Miron
CSIRO Livestock Industries, Australia
david.miron@csiro.au
I. V. Emelyanova
CSIRO Livestock Industries, Australia
irina.emelyanova@csiro.au
G. E. Donald
CSIRO Livestock Industries, Australia
graham.donald@csiro.au
G. M. Garner
Department Agriculture Fisheres and Forestry, Australia
graeme.garner@affa.gov.au

The modelling of livestock movements within Australia is of national importance for the purposes of the management and control of exotic disease spread, infrastructure development and the economic forecasting of livestock markets. In this paper an agent based model for the forecasting of livestock movements is presented. This models livestock movements from farm to farm through a saleyard. The decision of farmers to sell or buy cattle is often complex and involves many factors such as climate forecast, commodity prices, the type of farm enterprise, the number of animals available and associated off-shore effects. In this model the farm agent's intelligence is implemented using a fuzzy decision tree that utilises two of these factors. These two factors are the livestock price fetched at the last sale and the number of stock on the farm. On each iteration of the model farms choose either to buy, sell or abstain from the market thus creating an artificial supply and demand. The buyers and sellers then congregate at the saleyard where livestock are auctioned using a second price sealed bid. The price time series output by the model exhibits properties similar to those found in real livestock markets.

1 Introduction

The development of a national scale spatial simulation model for the modelling of livestock movements has been identified as a project of national importance by the Australian Department of Agriculture Fisheries and Forestry. Such a model would have applications in the development of policy and risk assessments surrounding exotic disease outbreaks, economic forecasting and infrastructure planning.

In this work an agent based modelling (ABM) approach is utilised for this purpose. The ABM approach is grounded in complex systems science and can be used to model non-linear systems such as those found in social and biological systems. For example, ABM techniques have been used in the development of models for the study of financial markets [1, 2, 11].

To utilise an ABM approach requires the identification of the agent types that are participating within the complex system and the development of the intelligence within the agents. The development of the intelligence within the agents is a key factor in determining realistic outputs for the model and is an area of active research [12].

The modelling of livestock movements can be achieved through the modelling of livestock markets as it is the latter that drives the former. Livestock markets are complex systems influenced by many factors. For example, rainfall, livestock prices, interest rates and feed prices all affect the decision making of stakeholders within the system which in turn influences the market.

For the purposes of this work the initial focus has been on beef cattle markets. Within the beef cattle sector the stakeholders are farmers, abattoirs, exporters, traders and feedlots. It is these stakeholders that are the agents within the model. However, in an effort to gauge the effectiveness of the ABM approach the model was simplified by considering only the farmer and saleyard agents. This simplification meant that only farm to farm movements through a saleyard were modelled.

Using this simplification and by implementing simple intelligence with farmer agents and a saleyard auction, it is shown that the price time series generated by the model has the same statistical and chaotic properties as those found in actual cattle markets.

This paper is divided into four sections. In §2 the livestock market is discussed and the models agents identified. In §3 the ABM that was implemented is described while in §4 results are presented. Conclusions and further work are given in §5.

2 The Market

A single trade between a buyer and a seller does not constitute a market. A market is created through a collection of trades between multiple buyers and sellers for a common commodity. The existence of a market is demonstrated through prices and trade volumes. The market operates under a number of different sets of rules such as the rules of supply and demand and the rules of trading.

Within cattle markets agents can buy or sell or as in the case of farmers take on the dual role of buyer and seller. The agents and their interactions are illustrated in Fig. 1.

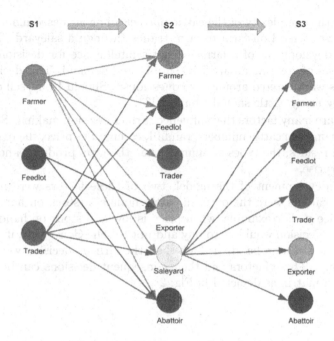

Figure 1: The daily trade cycle.

Fig. 1 illustrates the daily trade cycle of the agents participating in the cattle market. Farmers are the primary producers of livestock and depending upon their enterprise type may breed or fatten cattle for the purposes of sale. Traders are opportunists within the market only buying livestock to sell again in the short term. Exporters buy livestock for the purposes of export while abattoirs buy livestock for the purposes of slaughter. Feedlots are a special instance of a farmer in that they buy cattle for fattening before selling them on to an abattoir or an exporter. The saleyard has also been identified as an agent as it is a place where buyers and sellers congregate for the purposes of trading livestock through an auction.

Further, the daily trade cycle given in Fig. 1 is broken up into three stages, S1, S2 and S3. Stage S1 is made up of the agents that offer their cattle for sale while stage S2 is the group of agents, with the exception of saleyards, that would buy cattle. Again taking the example of farmers, they have the choice of selling directly to other agents such as other farmers and feedlots or offering their cattle for sale through a saleyard. It is the saleyard that necessitates the need for stage S3. Stage S3 is made up of agents that purchase cattle through a saleyard.

Immediately evident is the magnification role, in terms of the distribution of animals, that saleyards have within cattle markets.

3   The Agent Based Model

To reduce the complexity of the cattle market an agent based model was implemented that focused on farm to farm trades through a saleyard. This necessitated the development of a farmer agent's intelligence for decision making for the purposes of creating an artificial supply and demand. This decision making by farmers was centered around two questions: "Should I buy/sell cattle?" and if so, "How many cattle should I buy/sell?".

There are many factors that influence farmer decision making. Some of these factors are on farm cattle numbers, rainfall, commodity price, the region in which their farm resides, the types of animals that the farm produces and the type of farm enterprise.

In the development of the model, two of these factors were implemented. These two factors were the on farm cattle numbers (stock on hand) s and the market price p. For example, when price is low and stock on hand is low then the farmers decision would be to buy and not to sell. Conversely, if the price p is high and the stock on hand s is high then the farmers decision would be to sell and not buy. These factors and their consequent decisions can be represented by way of a matrix as depicted in Fig. 2:

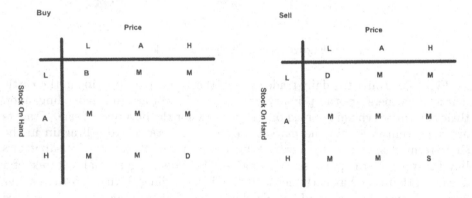

Figure 2: The sell/buy decision matrices.

The entries in the two matrices represent the cases don't sell/buy (D), buy (B), sell (S) and maybe buy/sell (M). The columns/rows of the matrix depict the status of each of the two factors while the values of the factors can either be low (L), average (A) or high (H).

These values low (L), average (A) or high (H) were represented by fuzzy variables. Fuzzy variables were first introduced by Zadeh [13]. The fuzzy variables in the decision matrices shown in Fig. 2 low, average and high were defined as:

$$\mu = \begin{cases} low = 0 \geq \text{price} \leq 0.9 \\ average = \text{price} \in \psi(1.0, 0.1) \\ high = 1.1 \geq \text{price} \leq 2.0 \end{cases} \tag{1}$$

Within equation 1 the function ψ is a trapezoidal function centered at 1.0 with endpoints at 0.9 and 1.1.

For the purposes of modelling, a saleyard in South East Queensland was chosen. The farms that participated in the market were those farms whose closest selling centre was the chosen saleyard. The stock on hand s for a farm i was calculated using the ratio of actual stock on hand $\hat{s}_i$ by average stock on hand $\bar{s}_i$. The value for the average stock on hand $\bar{s}_i$ was derived from Ag-Stats census data [3]. Thus, as a farm i bought and sold stock the fuzzy value of the stock on hand μ_i^s would change.

Initially, the current price ρ at the saleyard was set to $\rho = 1.0$ which yielded a fuzzy value $\mu^\rho = average$. The saleyard's current price ρ then fluctuated as a result of an auction process.

On each time step t farms were selected to buy and/or sell based on the outcome of applying the fuzzy decision matrices in Fig. 2. However, the decision matrices of Fig. 2 were modified to include a graded multiplication factor. The resulting matrices are given in Fig. 3.

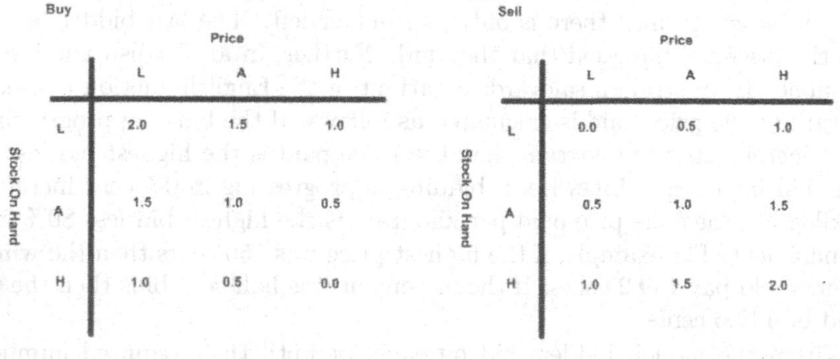

Figure 3: The multipliers for selection pressures of buying and selling.

The multiplication factors were used to compute a selection pressure for each farm i for each of the cases of buying and selling. The selection pressure for a selling farm γ_i^{sell} was calculated by first calculating the probability that farm i would sell. This probability $p_i(\text{sell})$ was the expected number of selling farms divided by the total number of farms. The expected number of selling farms was determined through expert estimates gathered by the Australian Department of Fisheries and Forestry (DAFF) [4].

The multiplier for farm i to sell m_i^{sell} was then applied to the probability of a farm selling $p_i(\text{sell})$ yielding:

$$\gamma_i^{\text{sell}} = m_i^{\text{sell}} \times p_i(\text{sell}).$$

A random number r_i where $r_i \in [0, 1)$ for farm i was then computed and applied to the function $g(r_i)$ where:

$$g(r_i) = \begin{cases} \text{participate} & \text{if } r_i < \gamma_i^{\text{sell}} \\ \text{abstain} & \text{otherwise} \end{cases} \tag{2}$$

If farm i was found to be a selling farm participating in the market then the number of stock needed to be calculated for sending to market. This was the actual stock on hand $0.1\hat{s}r_j$ where r_j was a random number and $r_j \in [0, 1)$.

The farms selection pressure for buying γ_i^{buy} was calculated in a similar manner and the decision made to buy using the same function $g(r_i)$. The number of stock to buy was also computed using the formula $0.1\hat{s}r_j$ where r_j is a random number and $r_j \in [0, 1)$.

Having completed the decision making process for buying and selling, the farms participating in the market began to buy and sell at auction. The actual auction technique implemented was a second price sealed bid. That is each buyer made one bid on the current lot and the bid was secret. That is no bidder knew what other bidders had bid. The winning bidder was the bidder with the highest bid but they only paid the losing bidder's price.

The second price sealed bid auction was used as it is weakly equivalent to an English auction [10]. In an English auction bidding for a good starts at low price and ascends until there is only one bidder left. The last bidder left then pays the price for the good that they bid. Further, in an English auction bids are public. In Australian saleyards a variant of the English auction is used. In this variant the price paid is calculated as follows. If the bids are progressing in 1 cent increments per kilogram then the price paid is the highest bid less 80% of the bid increment. Likewise, if bidding is progressing in 0.5 cent increments per kilogram then the pice paid per kilogram is the highest bid less 80% of the bid increment. For example, if the highest price was 150 cents then the winning bidder would pay 149.2 cents. If the increment was half cent bids then the price would be 149.6 cents.

Within the model, bidders bid on each lot until their required number of stock is purchased. After each round of bidding a market supply and demand factor was calculated. This modified the starting price of the bids for the lot. Actual bids were then modified by a choosing a random number from a Gaussian distribution centered at the "last lot price" with a variance of 0.005.

4 Results

A random output of the price time series from the model was taken. This time series was over a period of 1000 time steps. A time step could be considered a weekly sale. This time series was then tested for chaotic behaviour and compared against known cattle market chaos analysis to show that the cattle market created by the model has similar properties to a real cattle market.

The first step in the analysis of the model time series was to denoise the original signal. The wavelet shrinkage method of Donoho [5] was applied for this purpose. A shrinking threshold of 0.1 was selected for the signal denoising. Fig. 4 shows both the original signal (time series) on the left and the denoised signal on the right.

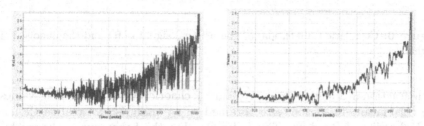

Figure 4: Time series of the original signal (left) and the denoised signal (right).

Fig. 5 shows the phase portraits for the time series of the original signal on the left and the denoised signal on the right.

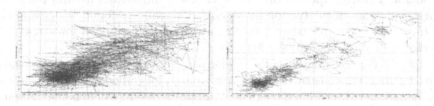

Figure 5: Phase portrait of the original signal (left) and the denoised signal (right).

From Fig. 5 it is evident that there are some similarities in distribution and density of trajectories in both two-dimensional phase space reconstructions. However, the trajectories of the denoised signal are smoother and free of random fluctuations. It may be seen that trajectories are quite mixed, change direction very quickly and build a complicated pattern. Thus, an assumption was made about chaotic behaviour of the model data. All the figures were produced using a freely available software package "Visual Recurrence Analysis 4.9" developed for topological analysis and qualitative assessment of time series [9].

The self-similarity of the model signal was tested by calculating correlation dimension introduced by Grassberger and Procaccia [6]. This measure helps quantify fractal features of the attractors. The graph of correlation dimensions for the reconstructed attractors of the original and denoised signals for the range of embedding dimensions $m = 1 - 17$ and time delay $\tau = 11$ is shown in Fig. 6.

From Fig. 6 (left) it can be seen that the curve for the original signal achieves its plateau at the value D_o 2.54. The value of the correlation dimension calculated for the denoised signal shown in Fig. 6 (right) was D_d 1.96 which is lower due to the reduction in the noise as compared to the original signal. This confirms the assumption about chaotic behaviour of the model data as attractors

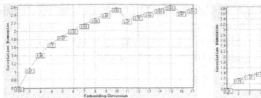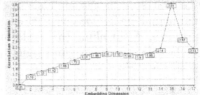

Figure 6: Correlation dimension for the original signal (left) and the denoised signal (right).

with fractional dimensions are typical of chaotic systems. Similar values for the correlation dimension were obtained in an analysis of a real cattle market by Kohzadi and Boyd [8]. This indicates that the behaviour of the model has produced chaotic properties similar to that of actual saleyards.

The fractal features of the model time series were also tested by the calculation of the Hurst exponent H [7]. It is known that a Hurst exponent of $H = 1$ results for smooth time series while a Hurst exponent of $H = 0.5$ characterises noise and a Hurst exponent of $0.5 <= H <= 1$ indicates a natural process. A Hurst exponent calculated for the denoised signal was found to have a value of $H = 0.62$ while the value of the Hurst exponent estimated for a real cattle market data was found to be a little lower at $H = 0.51$ [8].

To continue the analysis of the model time series a recurrence plot of the denoised signal was constructed. The recurrence plot indicated phase transitions and instationarities within the signal. Also, positive values of maximal Lyapunov exponents $L^l = 0.3$ and $L^d = 0.6$ calculated for the original and denoised signals respectively indicated the sensitive dependence on initial conditions. This suggests exponential divergence of nearby trajectories of the attractor and chaos.

5   Conclusions and Further Work

In this paper an agent based model has been put forward for the modelling of cattle movements. This model focussed on the movement of cattle from farm to farm through a saleyard. Using fuzzy decision tree for the decision making within the farm agents and an auction simulation it was shown that the model produced price time series that had characteristics similar to that of real saleyards. This has laid a strong foundation for further development of the model. It is envisaged that improving the agent decision making processes will be a key factor in improving the model outputs and is an area of active research.

Bibliography

[1] ARTHUR, W., "On learning and adaptation in the economy", *Tech. Rep. no. 92-07-038*, Santa fe Institute, (1992).

[2] ARTHUR, W., J. HOLLAND, B. Le BARON, R. PALMER, and P. TAYLER, "Asset pricing under endogenous expectations in an artificial stock market" (1996).

[3] "Australian bureau of statistics: Agstats census data" (2001).

[4] BECKETT, S., "Saleyard characteristics ausbeef region 6" (2006).

[5] DONOHO, D., "De-noising by soft-thresholding", *IEEE Trans. Inf. Theory* **41** (1994), 613–627.

[6] GRASSBERGER, P., and I. PROCACCIA, "Measuring the strangeness of strange attaractors", *Physica* **9D** (1983), 30–31.

[7] HURST, H., "Long-term storage capacity of reservoirs", *Transactions of American Society of Civil Engineering* **116** (1951), 770–799.

[8] KOHZADI, N., and M. BOYD, "Testing for chaos and nonlinear dynamics in cattle prices", *Canadian Journal of Agricultural Economics* **43** (1995), 475–484.

[9] KONONOV, E., "Visual recurrence analaysis", http://www.myjavaserver.com/~nonlinear/vra/download.html.

[10] KRISHNA, V., *Auction Theory*, Academic Press (2002).

[11] LeBARON, B., W. ARTHUR, and R. PALMER, "Time series properties of an artificial stock market", *Journal of Economic Dynamics & Control* **23** (1999), 1487–1516.

[12] PEREZ, P., *Complex Science for a Complex World: Exploring Human Ecosystems with Agents*, ANU Epress, Australia, (2006), ch. Chapter 3: Agents, Icons and Idols, pp. 27–56.

[13] ZADEH, L., "Fuzzy sets", *Information and Control* **8** (1965), 338–353.

Chapter 21

Random graph models of communication network topologies

Hannu Reittu and Ilkka Norros

first.last@vtt.fi

We consider a variant of so called power-law random graph. A sequence of expected degrees corresponds to a power-law degree distribution with finite mean and infinite variance. In previous works the asymptotic picture with number of nodes limiting to infinity has been considered. It was found that an interesting structure appears. It has resemblance with such graphs like the Internet graph. Some simulations have shown that a finite sized variant has similar properties as well. Here we investigate this case in more analytical fashion, and, with help of some simple lower bounds for large valued expectations of relevant random variables, we can shed some light into this issue. A new term, 'communication range random graph' is introduced to emphasize that some further restrictions are needed to have a relevant random graph model for a reasonable sized communication network, like the Internet. In this case a pleasant model is obtained, giving the opportunity to understand such networks on an intuitive level. This would be beneficial in order to understand, say, how a particular routing works in such networks.

1 Introduction

Since the pioneering work by three Faloutsos brothers, [1] and some other groups like that of Barabasi's [4], around the millennium, so called power-law graphs, with degree sequence obeying power-law distribution with finite mean and infinite variance, have attracted high interest by several authors. This degree sequence is argued to reflect some fundamental aspects of some communication networks and some other natural and technological networks. It appears that this concept can be turned into a mathematical object in several ways, with provable properties, see for instance [14, 2, 7, 8, 9, 13]. However, for the sake of tractability, asymptotic regime with growing number of nodes, $N \to \infty$, has been most popular. Some quite interesting results were obtained [11, 12]. For instance, it was found that a random graph variant of this model, where with a given degree sequence links are drawn as randomly as possible, produces a graph with some characteristic properties that has correspondence with such network as the Internet itself, at its autonomous systems level (AS-graph). A kind of 'soft-hierarchy' of large degree nodes is formed only due to combinatorial probabilities. It is convenient to categorize nodes with increasing degrees into 'tiers' with nodes in certain subsequent intervals of degrees. It is sufficient to consider approximately $\log \log N$ tiers. This can be done in a way that a node in a tier has at least one link to upper-laying next tier, with probability tending to 1, asymptotically. The hierarchical part of nodes was called the core. Thus very short ($\log \log N$) paths exist from the bottom to top of the core. It was also shown that even shorter paths are needed to find this core for almost any node in the same component. Thus, a $\log \log N$ asymptotical upper bound for distance in the giant component was established. These results with extensions were obtained independently by Chung and others, [2, 7], using some refined methods of random graph theory. However, we found our more elementary approach with the concept of the core also very useful giving an intuitive insight. Refined variants of the theory can be found in works of van der Hofstad and others, [8], showing, among other issues, that the $\log \log N$ upper bound is the best possible .

However, it was apparent from the proofs, that convergence can be very slow, involving such functions like $1/\log \log N$ with limit 0, that is approached only with 'unrealistically' large values of N, certainly unimaginable in the framework of communication networks. Some simulations indicated that in spite of this some reasonably sized graphs have properties that are similar to this asymptotic picture. That is why it is interesting to study this question in more details using an analytical approach. Here we do some first steps into this direction.

It is also interesting, that the asymptotic model indicates interesting consequences for such graphs robustness against targeted attacks against the top level nodes. Such graphs show good robustness against such failures, at least in the terms of the distance: the remaining tiers are able to maintain connectivity with the price of only an insignificant number of extra hops. However, here it is also important to know how valid these results are for reasonable graph sizes. Recently, possibilities of extending the basic model by 'redirecting' the links, have been discussed. Here it is possible that the asymptotic picture is 'non convergent', meaning that it tells nothing about the finite variant of the graph.

2 Model definition and asymptotic results

We consider a variant of power-law random graph, similar to one proposed by Chung and Lu [7], see also: [14, 13] . A natural number, $N > 0$, is the number of nodes in the graph. Nodes are labeled with natural numbers $1, 2, \cdots, N$. A node with label i, $1 \leq i \leq N$ has 'capacity', $\lambda_i = (N/i)^\alpha$, with real number α, $\frac{1}{2} < \alpha < 1$, which reflects the power-law degree sequence. For each possible unordered pair of nodes $\{i,j\}$, $i,j \in \{1,2,\cdots,N\}$, we associate the number of links between those nodes as a random variable $E_{i,j}$ with Poisson distribution, with expected value $(E_{i,j}) = \lambda_i\lambda_j / \sum_{i=1}^{N}\lambda_i$. All these random variables are considered as independent. In shorthand we write $E_{i,j} \cong Po(\lambda_i\lambda_j/L_N)$, with $L_N \equiv \sum_{i=1}^{N}\lambda_i$. Thus multiple links and self-loops are allowed. However, such artifacts are not too harmful, since the vast majority of these variables take values 0 or 1 in a large enough graph. The expected degree of node i, d_i, is thus $d_i = (\sum_{j=1}^{N}E_{i,j}) = \sum_{j=1}^{N}\lambda_i\lambda_j / \sum_{i=1}^{N}\lambda_i = \lambda_i$, due to a basic property for the sum of independent Poisson distributed random variables. Thus λ_i has the meaning of expected degree of node i.

Let us define the following sequence of functions:

$$\beta_0(N) = \frac{1}{\tau-1} + \frac{\epsilon(N)}{\tau-2} \tag{1}$$

$$\beta_j(N) = (\tau-2)\beta_{j-1}(N) + \epsilon(N), \quad j = 1,2,\cdots$$

with $\frac{1}{\tau-1} = \alpha$, $\epsilon(N) = l(N)/\log N$ and $l(\cdot)$ is a very slowly diverging function as its argument grows to infinity.

We define the 'upper layers' as

$$U_0 \equiv \{1\}, \quad U_j \equiv \{i : \lambda_i \geq N^{\beta_j(N)}\}, \quad j = 1,2,\cdots.$$

Provided that $l(\cdot)$ fulfills: $l(1) = 1$, $l(N)/\log\log\log N \to 0$ and $l(N)/\log\log\log\log N \to \infty$, we had the following result for the power-law graph described above: Let

$$k^* \equiv k^*(N) := \left\lceil \frac{\log\log N}{-\log(\tau-2)} \right\rceil,$$

where $\lceil\cdot\rceil$, denotes the least integer greater than or equal to its argument. Then the hop-count distance between two randomly chosen vertices of the giant component, which exists asymptotically almost surely (a.a.s.), is less than $2k^*(N)(1+o(1))$, a.a.s. We define the core, C, as the upper layer U_{k^*}. Later on this proposition was strengthened considerably, one such result being that this upper bound is tight [8]. However, such more detailed analysis is very involved and that is why we prefer to stay on the level of simple upper bounds, also in what follows. Roughly speaking, the idea of our proof of Theorem 2 is that the probability that a node i in any layer U_j has a link to the upper layer U_{j-1} with probability tending to 1 as $N \to \infty$ (we write this as: $U_j \ni i \leftrightarrow U_{j-1} \to 1$). Thus, it takes at most k^* hops to travel from the lowest layer U_k^* to the top degree node. Further, almost any node within the giant component has a path to some node in U_{k^*}, with a number of hops that is sub-linear with $\log\log N$. Thus, the upper bound follows. However, as we see in the next section, the convergence of probability to 1 can be very slow, say, $U_j \ni i \leftrightarrow U_{j-1} \geq 1 - \frac{c}{\log\log N}$, $c > 0$, a convergence rate that is practically 'unobservable' in our framework. Thus it is a relevant question whether this asymptotic picture tells anything about a graph with only a reasonably large number of nodes. Some simulations seem to indicate that the answer is positive, see e.g. [16, 12, 17]. However, we found that in order to find a corresponding above described layer structure in a finite model one must define function $l(\cdot)$ in some particular way, not prescribed by its asymptotic behavior only. In this paper our aim is to explain such circumstances and indicate a way how such finite sized random graphs can be analysed, and thus to make such random graph models more usable for modeling communication networks.

3   Analysis of 'communication range' graphs

It is easy to see that the cardinality of layer $j > 0$ is: $\mid U_j \mid = \lfloor N^{1-\beta_j(N)/\alpha} \rfloor = \lfloor N^{1-(\tau-1)\beta_j(N)} \rfloor$, where, $\lfloor \cdot \rfloor$ is the largest integer smaller or equal to the argument. Thus we have a lower bound for the sum of capacities in a layer $j > 0$:

$$V(U_j) \equiv \sum_{i \in U_j} \lambda_i \geq N^{\beta_j(N)} \left\lfloor N^{1-(\tau-1)\beta_j(N)} \right\rfloor \geq N^{1-(\tau-2)\beta_j(N)} - N^{\beta_j(N)} \equiv V_0(U_j)$$

and $V(U_0) = N^\alpha$. For L_N we have asymptotically linear scaling with N. Indeed, it is easy to see that $L_N \geq N^\alpha \int_0^N (1+x)^{-\alpha} dx \geq \frac{1}{1-\alpha}(N - N^\alpha) \geq \frac{c}{1-\alpha} N$, where c can be taken arbitrarily close to 1, provided that N is large enough. For instance, $c = 9/10$ is valid provided that $N > 10^{\frac{1}{1-\alpha}}$. However, notably these bounds are not uniform with α. For α, close to 1, we would have to choose a small value for c, for any reasonable N. This is a general trend here, since we must also fix the range of α more precisely, not just stating that $1/2 < \alpha < 1$, which was sufficient for the asymptotic analysis. This circumstance reflects the fact that the asymptotic regime is approached sensitively with respect to α. Our hypothesis is that for communication networks with N in reasonable range of thousands of nodes or tens thousands of nodes, it is necessary to have α in the lower half of the range $(0, 1)$ then the asymptotic range is reasonably close. Luckily, in the case of the Internet, this is a range of α that has been observed. We call this range of N and α, the *communication range*. Similarly we have:

$$V_0(U_j) \geq c_j N^{1-(\tau-2)\beta_j(N)}$$

with constants c_j, close to 1, provided N and τ are in the communication range. Within the same range, we can easily find a lower bound for the probability that a node in layer j has at least one link to layer $j - 1$:

$$U_j(N) \ni i \leftrightarrow U_{j-1}(N) = 1 - \exp(-\lambda_i V(U_{j-1})/L_N) \tag{2}$$
$$\geq 1 - \exp(-\frac{c_{j-1}}{c} e^{l(N)}) \equiv p_j.$$

This relation also shows the delicacy of the communication range, where we can approximate c_j/c by 1 — otherwise we would need a number far from 1, giving a big effect to the lower bound p_0. Say, if α is close to 1, this ratio would be a big number resulting in a very low probability. Notably, p_0 is also sensitive to the choice of function $l(\cdot)$. In the asymptotic sense these circumstances are irrelevant, since in any case $p_0 \to 1$ as $N \to \infty$, if only $l(N) \to \infty$.

We wish to show that in communication range, a particularly defined 'core' has a similar role as in the asymptotic graph. In particular, the $\log \log N$ scaling of distance is roughly valid. We show that the lower bound of expected number of nodes that have a link to a core node that has a path through the core hierarchy up to the top is large enough and suggests that the $\log \log N$ upper bound is valid within the communication range, for the vast majority of nodes.

Assume fixed N and take a natural number $x > 0$. The probability that a node in layer U_x, has a link to U_{x-1} is lower bounded by p_x. The probability that the same node has a path to U_{x-2}, through U_{x-1}, is lower bounded by $p_x p_{x-1}$. And so forth, probability that the same node has a path to the top node 1, going through at most x layers, is lower bounded by $p_x p_{x-1} \cdots p_1$. Denoting by c_x the minimal ratio $\frac{c_j}{c}$ in relation (2), in the corresponding range of j, and denoting

$$p_0 = 1 - \exp(-c_x e^{(3-\tau)l(N)}), \tag{3}$$

we find that the probability that a node in layer U_x has a path described above is lower bounded by p_0^x. Note that, in asymptotic range, p_0 tends to 1 (very slowly), however, within our finite range this is an important parameter affecting the quality of bounds.

Denote by $U_x' \subset U_x$ the nodes in layer U_x having a valid path with upmost x hops within the subsequent layers to the top node 1. As a result we have:

$$| U_x' | \geq p_0^x | U_x | \tag{4}$$
$$V(U_x') \geq N^{\beta_x} p_0^x | U_x | .$$

Denote by N_x the nodes that are in U_x' or have a link to a node in it. The probability that a node outside U_x' has link to it is lower bounded by $1 - \exp(-\frac{V(U_x')}{L_N}) \geq \frac{1}{2}\frac{V(U_x')}{L_N}$. That is why, for the conditional expectation, we have:

$$(| N_x ||| U_x' |) \geq \frac{N}{2}\frac{V(U_x')}{L_N}.$$

Therefore, according to (4),

$$(| N_x |) \geq \frac{N^{\beta_x} N^1}{2L_N} p_0^x | U_x | . \tag{5}$$

The task is to maximise the bound (5), in a way that x is not too large. It appears that asymptotically we end up with the setting that is in line with Theorem 2. However, it is also possible to find a 'setting' of $l(\cdot)$ and x that corresponds to asymptotic-like behavior in the communication range. To get a qualitative picture, we make simplifications in relation (5), assuming all constants, that are close to 1, equal to one. As a result we get an approximate lower bound for $(|N_x|) \geq s(x, l)$:

$$s(x, l) \approx N^{1 - \frac{(\tau - 2)^{x+1}}{\tau - 1}} (1 - \exp(-m))^x m^{-\frac{\tau - 2}{3 - \tau}}, \quad m \equiv \exp((3 - \tau)l). \quad (6)$$

The maximum is found as solution of equations $\frac{\partial s(x,l)}{\partial x} = 0$, $\frac{\partial s(x,l)}{\partial l} = 0$, and by comparing values of the function in the closest integer arguments. Although the equations are not solvable in closed form, the first one yields the relation:

$$\frac{\partial s(x, l)}{\partial x} = 0 \Leftrightarrow x = \frac{\log \log N}{-\log(\tau - 2)} + \log\left(\frac{\log(1/(1 - \exp(-m)))}{-\log(\tau - 2)}\right) - 1, \quad (7)$$

The first term is analogous to the one in Theorem 2, the next one depends on the choice of l through $m(l)$, which should be found from the second equation. However, we can see the asymptotical regime from these equations. Indeed, we see that the relevant factor of s, with respect to argument l, is asymptotically $m^{-\frac{\tau - 2}{3 - \tau}} \exp\left(\frac{-x}{\exp(m)}\right) \sim m^{-\frac{\tau - 2}{3 - \tau}} \exp\left(\frac{\log \log N}{\log(\tau - 2) \exp(m)}\right)$, where we took into account the equation (7). The second term suggests that m should be an increasing function, at least threefold logarithm, and l should be no slower than fourfold logarithm. The first term suggests that it should not be too fast, in this respect the lowest possible would be the best, and gives the maximum. However, as suggested by the Theorem 2, the leading term is indifferent with respect to this range. Indeed, if we make the corresponding substitution to (7), we see that the leading term is just k^*.

If we substitute these asymptotic estimates, as arguments to (6), we find

$$s(x, l) \propto N \exp(-(\tau - 2)l(N)), \quad (8)$$

which is only slightly lower than N. Thus, the factor $\exp(-(\tau - 2)l(N))$ has the meaning of lower bound expected 'density' of neighbors of the core nodes, with valid paths to the top node. This density is almost constant, as a function of N, between $1/\log \log \log N$ and $1/\log \log N$. In the communication range, one would guess that the best choice would be $1/\log \log \log N$. Numerical calculations seems to support this, see Figure 1. Our next plot, in Figure 2, indicates that in this range the choice of this function has some effect. By taking for the x its optimal value x^*, we see that the function $s(x^*, l)$, see 6, has a visible maximum. This numerics showed also that in a very wide range of N, from 100 to 10^{10}, a rather constant fraction of nodes, in lower bound expectation, are neighbors of core with valid paths to the top. In our case with $\tau = 2.5$, this was around 20 percent. As a result, a random node is able to find a node in the core that has a path to its top, with a moderate sized search: approximately every fifth node is of this type.

This kind of 'quasi-stationarity' or extremely low dependence on N should be good news, since it simplifies the usability of such models. For instance, $s(x, l)$ and l can be taken as constant parameters hardly changing in any reasonable range. We also see that the height of the core is almost constant, and its major term is a function of type $k^*(N)$. One drawback is that we have only lower bound

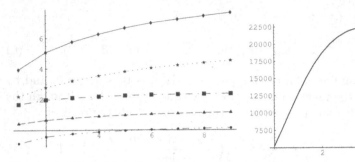

Figure 1: Plot of functions, calculated in points $N = 10^{k+1}, k = 1, 2, \cdots, 9$: $x(N)$, $\frac{1}{\log 1/(\tau-2)} \log\log N$, $l(N)$, $\log\log\log N$ and $\log\log\log\log N$, listed from top to down.

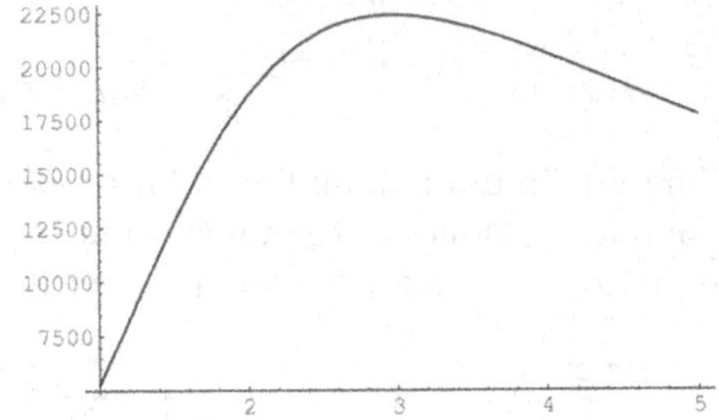

Figure 2: Plot of function $s(x^*, l)$, with $N = 10^5$ and with fixed first argument with value x^*, that it takes at maximum.

type results and some unrigorous estimates were done. However, it is quite likely that a thorough analysis will not reveal any substantial new features, although it is mandatory to check it. It would be interesting to compare this approach with a 'conceptual model' of the Internet, called the 'Jellyfish', [15].

Acknowledgments This work was financially supported by EU-project Net-Refound, Project Number: 034413.

Bibliography

[1] FALOUTSOS, M. FALOUTSOS, P. AND FALOUTSOS, C. "On the power-law relationships of the Internet topology". In *Proc. of the ACM/SIGCOMM'99*(1999), pp. 251–261.

[2] AIELLO, W., CHUNG, F. AND LU, L. "A random graph model for massive graphs". In *Proc. of the 32nd Annual ACM Symposium on Theory of Computing*, (2000) pp. 171–180.

[3] BARABÁSI, A.-L. AND ALBERT, R. "Emergence of scaling in random networks". *Science* **286**,(1999), 509–512.

[4] BARABÁSI, A.-L. AND ALBERT, R. "Statistical mechanics of complex networks". *Rev. Mod. Phys.* **74**,(2002), 47–97.

[5] BOLLOBÁS, B. AND RIORDAN, O. "Coupling scale-free and classical random graphs". *Internet Mathematics* **1**,(2003), 215–225.

[6] BOLLOBÁS, B. AND RIORDAN, O. "The diameter of a scale-free random graph". *Combinatorica*, **4**,(2004), 5–34.

[7] CHUNG, F. AND LU, L. "The average distance in a random graph with given expected degrees" (full version). *Internet Mathematics*, **1**,(2003), 91–114.

[8] VAN DER HOFSTAD, R., HOOGHIEMSTRA, G. AND ZNAMENSKI, D. "Distances in random graphs with finite mean and infinite variance degrees". Preprint, submitted for publication (2005).

[9] HOOGHIEMSTRA, G. AND VAN MIEGHEM, P. "On the mean distance in scale-free graphs". *Methodology and Computing in Applied Probability*. To appear (2005).

[10] NEWMAN, M.E.J., STROGATZ, S.H. AND WATTS, D.J. "Random graphs with arbitary degree distribution and their applications". *Phys. Rev. E* **64,**(2001), 026118.

[11] REITTU, Hannu, and NORROS, Ilkka . "On the effect of very large nodes in Internet graphs". In *Globecom'02*(2002), Taipei, Taiwan.

[12] REITTU, Hannu, and NORROS, Ilkka . "On the power-law random graph model of massive data networks". *Performance Evaluation*(2004), 3–23.

[13] NORROS, Ilkka and REITTU, Hannu . "On a conditionally Poissonian graph process", *Advances in Applied Probability* **38,**(2006), 59–75.

[14] BOLLOBAS, B. JANSON S. AND RIORDAN O. "The phase transition in inhomogeneous random graphs" *Random Structures and Algorithms* **31(3),**(2007), 3–122.

[15] SIGANOS, G. TAURO, S.L. and FALOUTSOS, M. "Jellyfish: A conceptual model for the AS Internet topology" *Journal of Communications and Networks* **8(3)**(2006), 339–350.

[16] TANGMUNARUNKIT, H. GOVINDAN, R. JAMIN, S. SHENKER, S.and WILLINGER, W.. "Network topology generators: degree based vs. structural". In *Proc. of the SIGCOMM'02*(2002).

[17] NORROS, Ilkka and REITTU Hannu . "Architectural Features of the Power-Law Random Graph Model of Internet: Notes on Soft Hierarchy, Vulnerability and Multicasting". In *Proc. of the ITC-18*(Sep. 2003).

221

Chapter 22

Using RDF to Model the Structure and Process of Systems

Marko A. Rodriguez[1,2*]
Jennifer H. Watkins[2*]
Johan Bollen[1,3*]
Carlos Gershenson[4,5,6*]
[1]Los Alamos National Laboratory
[2]AT&T Interactive
[3]Indiana University
[4]New England Complex Systems Institute
[5]Vrije Universiteit Brussel
[6]Universidad Nacional Autónoma de México

Many systems can be described in terms of networks of discrete elements and their various relationships to one another. A semantic network, or multi-relational network, is a directed labeled graph consisting of a heterogeneous set of entities connected by a heterogeneous set of relationships. Semantic networks serve as a promising general-purpose modeling substrate for complex systems. Various standardized formats and tools are now available to support practical, large-scale semantic network models. First, the Resource Description Framework (RDF) offers a standardized semantic network data model that can be further formalized by ontology modeling languages such as RDF Schema (RDFS) and the Web Ontology Language (OWL). Second, the recent introduction of highly performant triple-stores (i.e. semantic network databases) allows semantic network models on the order of 10^9 edges to be efficiently stored and manipulated. RDF and its related technologies are currently used extensively in the domains of computer science, digital library science, and the biological sciences. This article will provide an introduction to RDF/RDFS/OWL and an examination of its

---

*Current affiliations.

suitability to model discrete element complex systems.

1 Introduction

The figurehead of the Semantic Web initiative, Tim Berners-Lee, describes the Semantic Web as

> ... an extension of the current web in which information is given well-defined meaning, better enabling computers and people to work in cooperation [2].

However, Berners-Lee's definition assumes an application space that is specific to the "web" and to the interaction between humans and machines. More generally, the Semantic Web is actually a conglomeration of standards and technologies that can be used in various disparate application spaces. The Semantic Web is simply a highly-distributed, standardized semantic network (i.e. directed labeled network) data model and a set of tools to operate on that data model. With respect to the purpose of this article, the Semantic Web and its associated technologies can be leveraged to model and manipulate any system that can be represented as a heterogeneous set of discrete elements connected to one another by a set of heterogeneous relationships whether those elements are web pages, automata, cells, people, cities, etc. This article will introduce complexity science researchers to a collection of standards designed for modeling the heterogeneous relationships that compose systems and technologies that support large-scale data sets on the order to 10^9 edges.

This article has the following outline. Section 2 presents a review of the Resource Description Framework (RDF). RDF is the standardized data model for representing a semantic network and is the foundational technology of the Semantic Web. Section 3 presents a review of both RDF Schema (RDFS) and the Web Ontology Language (OWL). RDFS and OWL are languages for abstractly defining the topological features of an RDF network and are analogous, in some ways, to the database schemas of relational databases (e.g. MySQL and Oracle). Section 4 presents a review of triple-store technology and its similarities and differences with the relational database. Finally, Section 5 presents the semantic network programming language Neno and the RDF virtual machine Fhat.

2 The Resource Description Framework

The Resource Description Framework (RDF) is a standardized data model for representing a semantic network [5]. RDF is not a syntax (i.e. data format). There exist various RDF syntaxes and depending on the application space one syntax may be more appropriate than another. An RDF-based semantic network is called an RDF network. An RDF network differs from the directed network of common knowledge because the edges in the network are qualified. For instance, in a directed network, an edge is represented as an ordered pair

(i,j). This relationship states that i is related to j by some unspecified type of relationship. Because edges are not qualified, all edges have a homogenous meaning in a directed network (e.g. a coauthorship network, a friendship network, a transportation network). On the other hand, in an RDF network, edges are qualified such that a relationship is represented by an ordered triple $\langle i, \omega, j \rangle$. A triple can be interpreted as a statement composed of a subject, a predicate, and an object. The subject i is related to the object j by the predicate ω. For instance, a scholarly network can be represented as an RDF network where an article cites an article, an author collaborates with an author, and an author is affiliated with an institution. Because edges are qualified, a heterogeneous set of elements can interact in multiple different ways within the same RDF network representation. It is the labeled edge that makes the Semantic Web and the semantic network, in general, an appropriate data model for systems that require this level of description.

In an RDF network, elements (i.e. vertices, nodes) are called resources and resources are identified by Uniform Resource Identifiers (URI) [1]. The purpose of the URI is to provide a standardized, globally-unique naming convention for identifying any type of resource, where a "resource" can be anything (e.g. physical, virtual, conceptual, etc.). The URI allows every vertex and edge label in a semantic network to be uniquely identified such that RDF networks from disparate organizations can be unioned to form larger, and perhaps more complete, models. The Semantic Web can span institutional boundaries to support a world-scale model. The generic syntax for a URI is

`<scheme name> : <hierarchical part> [ # <fragment> ]`

Examples of entities that can be denoted by a URI include:

- a physical object (e.g. `http://www.lanl.gov/people#marko`)
- a physical component (e.g. `http://www.lanl.gov/people#markos_arm`)
- a virtual object (e.g. `http://www.lanl.gov/index.html`)
- an abstract class (e.g. `http://www.lanl.gov/people#Human`).

Even though each of the URIs presented above have an `http` schema name, only one is a Uniform Resource Locator (URL) [9] of popular knowledge: namely, `http://www.lanl.gov/index.html`. The URL is a subclass of the URI. The URL is an address to a particular harvestable resource. While URIs can point to harvestable resources, in general, it is best to think of the URI as an address (i.e. pointer) to a particular concept. With respects to the previously presented URIs, Marko, his arm, and the class of humans are all concepts that are uniquely identified by some prescribed globally-unique URI.

Along with URI resources, RDF supports the concept of a literal. Example literals include the integer 1, the string "marko", the float (or double) 1.034, the date 2007-11-30, etc. Refer to the XML Schema and Datatypes (XSD) specification for the complete classification of literals [3].

If U is the set of all URIs and L is the set of all literals, then an RDF network (or the Semantic Web in general) can be formally defined as[1]

$$G \subseteq \langle U \times U \times (U \cup L) \rangle. \tag{1}$$

To ease readability and creation, schema and hierarchies are usually prefixed (i.e. abbreviated). For example, in the following two triples, lanl is the prefix for http://www.lanl.gov/people#:

```
<lanl:marko, lanl:worksWith, lanl:jhw>
<lanl:marko, lanl:hasBodyPart, lanl:markos_arm>
```

These triples are diagrammed in Figure 1. The union of all RDF triples is the Semantic Web.

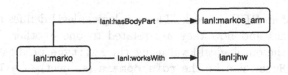

Figure 1: Two RDF triples as an RDF network.

The benefit of RDF, and perhaps what is not generally appreciated, is that with RDF it is possible to represent anything in relation to anything by any type of qualified relationship. In many cases, this generality can lead to an uncontrolled soup of relationships; however, thanks to ontology languages such as RDFS and OWL, it is possible to formally constrain the topological features of an RDF network and thus, subsets of the larger Semantic Web.

3 The RDF Schema and Web Ontology Language

The Resource Description Framework and Schema (RDFS) [4] and the Web Ontology Language (OWL) [6] are both RDF languages used to abstractly define resources in an RDF network. RDFS is simpler than OWL and is useful for creating class hierarchies and for specifying how instances of those classes can relate to one another. It provides three important constructs: rdfs:domain, rdfs:range, and rdfs:subClassOf[2]. While other constructs exist, these three tend to be the most frequently used when developing an RDFS ontology. Figure 2 provides an example of how these constructs are used. With RDFS (and OWL), there is a sharp distinction between the ontological- and

---

[1]Note that there also exists the concept of a blank node (i.e. anonymous node). Blank nodes are important for creating n-ary relationships in RDF networks. Please refer to the official RDF specification for more information on the role of blank nodes.

[2]rdfs is a prefix for http://www.w3.org/2000/01/rdf-schema#

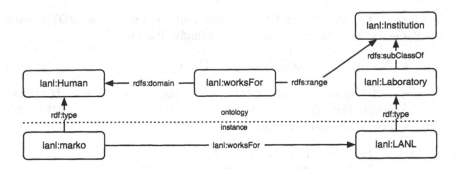

Figure 2: The relationship between an instance and its ontology.

instance-level of an RDF network. The ontological-level defines abstract classes (e.g. `lanl:Human`) and how they are related to one another. The instance-level is tied to the ontological-level using the `rdf:type` predicate[3]. For example, any `lanl:Human` can be the `rdfs:domain` (subject) of a `lanl:worksFor` triple that has a `lanl:Institution` as its `rdfs:range` (object). Note that the `lanl:Laboratory` is an `rdfs:subClassOf` a `lanl:Institution`. According to the property of subsumption in RDFS reasoning, subclasses inherit their parent class restrictions. Thus, `lanl:marko` can have a `lanl:worksFor` relationship with `lanl:LANL`. Note that RDFS is not intended to constrain relationships, but instead to infer new relationships based on restrictions. For instance, if `lanl:marko lanl:worksFor` some other organization denoted X, it is inferred that that X is an `rdf:type` of `lanl:Institution`. While this is not intuitive for those familiar with constraint-based database schemas, such inferencing of new relationships is the norm in the RDFS and OWL world.

Beyond the previously presented RDFS constructs, OWL has one primary construct that is used repeatedly: `owl:Restriction`[4]. Example `owl:Restrictions` include, but are note limited to, `owl:maxCardinality`, `owl:minCardinality`, `owl:cardinality`, `owl:hasValue`, etc. With OWL, it is possible to state that a `lanl:Human` can work for no more than 1 `lanl:Institution`. In such cases, the `owl:maxCardinality` restriction would be specified on the `lanl:worksFor` predicate. If there exist the triples

```
<lanl:marko, lanl:worksFor, lanl:LANL>
<lanl:marko, lanl:worksFor, lanl:LosAlamos>,
```

an OWL reasoner would assume that `lanl:LANL` and `lanl:LosAlamos` are the same entity. This reasoning is due to the cardinality restriction on the `lanl:worksFor` predicate.

There are two popular tools for creating RDFS and OWL ontologies:

---

[3]`rdf` is a prefix for http://www.w3.org/1999/02/22-rdf-syntax-ns#
[4]`owl` is a prefix for http://www.w3.org/2002/07/owl#

226

Protégé[5] (open source) and Top Braid Composer[6] (proprietary).

4 The Triple-Store

There are many ways in which RDF networks are stored and distributed. In the simple situation, an RDF network is encoded in one of the many RDF syntaxes and made available through a web server (i.e. as a web document). In other situations, where RDF networks are large, a triple-store is used. A triple-store is to an RDF network what a relational database is to a data table. Other names for triple-stores include semantic repository, RDF store, graph store, RDF database. There are many different propriety and open-source triple-store providers. The most popular proprietary solutions include AllegroGraph[7], Oracle RDF Spatial[8] and the OWLIM semantic repository[9]. The most popular open-source solution is Open Sesame[10].

The primary interface to a triple-store is SPARQL [7]. SPARQL is analogous to the relational database query language SQL. However, SPARQL is perhaps more similar to the query model employed by logic languages such as Prolog. The example query

```
SELECT ?x
  WHERE { ?x <lanl:worksWith> <lanl:jhw> . }
```

returns all resources that work with `lanl:jhw`. The variable `?x` is a binding variable that must hold true for the duration for the query. A more complicated example is

```
SELECT ?x ?y
  WHERE {
    ?x <lanl:worksWith> ?y .
    ?x <rdf:type> <lanl:Human> .
    ?y <rdf:type> <lanl:Human> .
    ?y <lanl:worksFor> <lanl:LANL> .
    ?x <lanl:worksFor> <necsi:NECSI> . }
```

The above query returns all collaborators such that one collaborator works for the Los Alamos National Laboratory (LANL) and the other collaborator works for the New England Complex Systems Institute (NECSI). An example return would be

```
---------------------------------
|    ?x      |      ?y      |
---------------------------------
| lanl:marko   | necsi:carlos |
| lanl:jhw     | necsi:carlos |
| lanl:jbollen | necsi:carlos |
---------------------------------
```

---

[5]Protégé available at: http://protege.stanford.edu/
[6]Top Braid Composer available at: http://www.topbraidcomposer.com/
[7]AllegroGraph available at: http://www.franz.com/products/allegrograph/
[8]Oracle RDF Spatial available at: http://www.oracle.com/technology/tech/semantic_technolog:
[9]OWLIM available at: http://www.ontotext.com/owlim/
[10]Open Sesame available at: http://www.openrdf.org/

The previous query would require a complex joining of tables in the relational database model to yield the same information. Unlike the relational database index, the triple-store index is optimized for such semantic network queries (i.e. multi-relational queries). The triple-store a useful tool for storing, querying, and manipulating an RDF network.

5 A Semantic Network Programming Language and an RDF Virtual Machine

Neno/Fhat is a semantic network programming language and RDF virtual machine (RVM) specification [8]. Neno is an object-oriented language similar to C++ and Java. However, instead of Neno code compiling down to machine code or Java byte-code, Neno compiles to Fhat triple-code. An example Neno class is

```
owl:Thing lanl:Human {
  lanl:Institution lanl:worksFor[0..1];

  xsd:nil lanl:quit(lanl:Institution x) {
    this.worksFor =- x;
  }
}
```

The above code defines the class `lanl:Human`. Any instance of `lanl:Human` can have either 0 or 1 `lanl:worksFor` relationships (i.e. `owl:maxCardinality` of 1). Furthermore, when the method `lanl:quit` is executed, it will destroy any `lanl:worksFor` triple from that `lanl:Human` instance to the provided `lanl:Institution x`.

Fhat is a virtual machine encoded in an RDF network and processes Fhat triple-code. This means that a Fhat's program counter, operand stack, variable frames, etc., are RDF sub-netwoks. Figure 3 denotes a Fhat processor (**A**) processing Neno triple-code (**B**) and other RDF data (**C**).

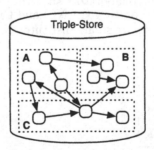

Figure 3: The Fhat RVM and Neno triple-code commingle with other RDF data.

With Neno it is possible to represent both the system model and its algorithmic processes in a single RDF network. Furthermore with Fhat, it is possible

228

to include the virtual machine that executes those algorithms in the same substrate. Given that the Semantic Web is a distributed data structure, where sub-networks of the larger Semantic Web RDF network exist in different triple-stores or RDF documents around the world, it is possible to leverage Neno/Fhat to allow for distributed computing across these various data sets. If a particular model exists at domain X and a researcher located at domain Y needs to utilize that model for a computation, it is not necessary for the researcher at domain Y to download the data set from X. Instead, a Fhat processor and associated Neno code can move to domain X to utilize the data and return with results. In Neno/Fhat, the data doesn't move to the process, the process moves to the data.

6 Conclusion

This article presented a review of the standards and technologies associated with the Semantic Web that can be used for complex systems modeling. The World Wide Web provides a common, standardized substrate whereby researchers can easily publish and distribute documents (e.g. web pages, scholarly articles, etc.). Now with the Semantic Web, researchers can easily publish and distribute models and processes (e.g. data sets, algorithms, computing machines, etc.).

Bibliography

[1] BERNERS-LEE, Tim, , R. FIELDING, Day SOFTWARE, L. MASINTER, and Adobe SYSTEMS, "Uniform Resource Identifier (URI): Generic Syntax" (January 2005).

[2] BERNERS-LEE, Tim, James A. HENDLER, and Ora LASSILA, "The Semantic Web", *Scientific American* (May 2001), 34–43.

[3] BIRON, Paul V., and Ashok MALHOTRA, "XML schema part 2: Datatypes second edition", *Tech. Rep. no.*, World Wide Web Consortium, (2004).

[4] BRICKLEY, Dan, and R.V. GUHA, "RDF vocabulary description language 1.0: RDF schema", *Tech. Rep. no.*, World Wide Web Consortium, (2004).

[5] MANOLA, Frank, and Eric MILLER, "RDF primer: W3C recommendation" (February 2004).

[6] McGUINNESS, Deborah L., and Frank van HARMELEN, "OWL web ontology language overview" (February 2004).

[7] PRUD'HOMMEAUX, Eric, and Andy SEABORNE, "SPARQL query language for RDF", *Tech. Rep. no.*, World Wide Web Consortium, (October 2004).

[8] RODRIGUEZ, Marko A., "General-purpose computing on a semantic network substrate", *Tech. Rep. no. LA-UR-07-2885*, Los Alamos National Laboratory, (2007).

[9] W3C/IETF, "URIs, URLs, and URNs: Clarifications and recommendations 1.0" (September 2001).

Chapter 23

How do agents represent?

Alex Ryan
DSTO, Australia
alex.ryan@dsto.defence.gov.au

Representation is inherent to the concept of an agent, but its importance in complex systems has not yet been widely recognised. In this paper I introduce Peirce's theory of signs, which facilitates a definition of representation in general. In summary, representation means that for some agent, a model is used to stand in for another entity in a way that shapes the behaviour of the agent with respect to that entity. Representation in general is then related to the theories of representation that have developed within different disciplines. I compare theories of representation from metaphysics, military theory and systems theory. Additional complications arise in explaining the special case of mental representations, which is the focus of cognitive science. I consider the dominant theory of cognition – that the brain is a representational device – as well as the sceptical anti-representational response. Finally, I argue that representation distinguishes agents from non-representational objects: agents are objects capable of representation.

1.1 Introduction

Representation is an essential concept for understanding the behaviour of agents in a complex system. Consider traders in a stock exchange market as agents. If every agent has unmediated access to the value of a company (including its exact future profits discounted to present value), then the market cannot exist, since shareholders would only be willing to sell above this value, a price no rational buyer would pay[1]. Only when partial information on value is allowed

---

[1]One might expect trades to be made exactly at the value of the stock. However, once a financial or time cost is included no rational buyer can exist. Why would an agent buy shares that never increase in real value and incur an exit fee?

and different agents have access to different information is it possible to predict the formation of a market. In this case, each agent must construct a model representing the perceived value of a company. By communicating, agents can modify their models to take into account the representations other agents in their social network have constructed. Because there is a benefit in being connected to agents who are better at predicting future value, some agents may specialise in developing predictive models and charging other agents for access to their expectations (such as financial advisors). Markets would not exist if there were not differences between agents in their representations. Variety in representation allows the simultaneous existence of buyers and sellers, as well as the potential for a secondary market based on constructing representations and selling advice. Even though this is quite obvious, imperfect information, bounds on rationality, and consequently the need for constructing representations did not feature in the theories and models of classical economics.

It turns out that an account of representation is just as important in understanding the role of the discipline of complex systems, as for understanding the behaviour of agents within a complex system. This is because the systems approach is a way of representing the world. When this is overlooked, systems applications may be blind to the limitations of the representations they employ. This discussion of representation is intended to be interpreted on two levels. On one level, when an analyst uses a complex systems approach, they invariably construct systems representations. On another level, when the system contains agents that also represent their environment, this must be accounted for in any model of the system.

Section 1.2 makes the metaphysical assumptions of this paper explicit. Then in Section 1.3, Peirce's theory of signs is used as a basis for a theory of representation in general. When agents represent their environment, they may use either external or internal models. Section 1.4 surveys accounts of external representation across several disciplines, while Section 1.5 surveys internal representation, which has been discussed mostly in philosophy of mind and cognitive science. This paper concludes by defining 'agent' in Section 1.6, which demonstrates the strong link between agency and representation.

1.2   Metaphysical Assumptions

Before representation is discussed in detail, it is prudent to make the metaphysical assumptions of this paper explicit. The metaphysical position I will adhere to is known as physicalism, the view that there are no kinds of things other than physical things. In particular, I assume that the relationship between macroscopic and microscopic phenomena is one of supervenience. The Stanford Encyclopedia of Philosophy offers the following definition:

Definition 1 (Supervenience) *A set of properties A supervenes upon another set B just in case no two things can differ with respect to A-properties without also differing with respect to their B-properties [40].*

Supervenience, along with physicalism, entails that in principle, all of the book-keeping regarding forces can be accounted for in purely physical terms between arbitrarily small entities, when the set B is taken to be the properties of fundamental physics. This is because every time there is a change in a macro level property, there must be a corresponding change in the micro level properties. That the physcial forces fully account for the dynamics at the micro level tells us little about what physical predictions *mean*. Semantics is always relative to an agent's subjective experience of the world, a concept which does not feature in, and cannot be fully explained by, the elementary particles and fundamental forces of physics. First-person experience is just one example of an emergent property, the general reason why descriptions at other levels cannot be eliminated. I will assume that forces in chemistry, biology, psychology and sociology do not add anything to the physical: that the laws of physics are *conservative*. This is consistent with Anderson's [3] twin assertions that all ordinary matter obeys simple electrodynamics and quantum theory, but that "the ability to reduce everything to simple fundamental laws does not imply that ability to start from those laws and reconstruct the universe". This assumption can be argued with, but it cannot be proved either way. I assume supervenience regarding the relationship between macro and micro phenomena because to do otherwise is to place some entities outside the domain of scientific explanation, and it is difficult to see what is achieved by doing so. Descartes' [21] non-physical mind that provided the basis for substance dualism in Meditations VI, and Bergson's [8] *elan vital* that animated the evolution and development of organisms, are examples of non-physical entities that have been postulated in science, and history suggests both acted as barriers to progress. Consequently, I only consider representations that supervene on the physical as meaningful.

1.3 Representation in General

Things don't mean: we construct meaning using representational systems – concepts and signs.

Stuart Hall

There are a number of reasons why unmediated interaction with the world can be undesirable. Some entities are distinctly unfriendly, others are inaccessible, and sometimes the process of interaction is too costly or time consuming. In order to understand anything about a solar flare on the surface of the Sun, mediated access, via the construction of models, is necessary to avoid the undesirable consequences of unmediated contact. A model acts as a representation because it *stands in* for unmediated interaction with the system of interest. Other situations where representations stand in for unmediated interaction include predicting properties of previously unrealised configurations; designing artifacts that do not exist; facilitating comparison of structural similarities between apparently dissimilar phenomena; and generalising knowledge to apply beyond a single entity at a single moment in time[2]. As will be discussed in Section 1.5.1, the dominant theory of human cognition assumes that the mind is a representational device, and that the brain has representational content.

In counterpoint to the important and varied roles of representation, there exists little formal work on representation in general. What are the necessary and sufficient conditions for representation to occur? What kinds of representations exist? One such general theory was proposed by Peirce, which he named the theory of signs or "semiotics". However, because much of his work was misplaced and posthumously edited non-chronologically into highly fragmented volumes, and since Peirce's unique and subtle philosophy requires explication before the finer points of his theory of signs can be appreciated, it remains under-utilised as a general theory of representation. Fortunately, Von Eckardt [63, p. 143-159] has performed considerable work to situate Peirce's theory of signs as a foundation for the more specialised debate on mental representation in cognitive science. I will draw heavily on Von Eckardt's interpretation of Peirce, since it is better oriented towards contemporary concerns in the theory of representation. Unlike Peirce or Von Eckardt, my interests apply to the field of complex systems, and so I will abuse the semiotic and cognitive science terminology by translating it into more general language.

For Peirce, representation was an irreducible triadic relation between objects, signs and interpretants. This implies that something is a sign only if it is a sign *of* an object *with respect to* an interpretant [63, p. 145]. It also implies that the representation relation cannot be decomposed into diadic relations between entities, objects and signs. According to Peirce, representation can only be fully understood by considering the three components of the triadic relation simultaneously.

In the triadic relation, the sign is a token that signifies the object. The more general term I will substitute for sign is model. Peirce's object is already quite general: it may be abstract or concrete, a singular object or a set of objects (a complex object). However, since objects do not in fact need to be objective (or concrete), I will use the more general term entity. The interpretant exists in the mind of the interpreter for whom the sign is a sign [63, p. 148]. Whereas Peirce and Von Eckardt limit their attention to human interpreters, I will generalise this to consider agents. The cost of this generalisation is that the precise nature of interpretants cannot be specified in a way that applies to all agents. Consequently, I can only induce the existence of an interpretant by an observable effect on the agent's behaviour. The triadic relation between entity E, model M and interpretant I is illustrated in Figure 1.1.

Von Eckardt [63, p. 158] summarises the value of Peirce's theory of signs for understanding representation in general as follows:

1. Peirce's distinction between a representation and a representation bearer;

2. His insistence that something can be a full-blown representation only if it is both grounded and interpreted;

---

[2]This list paraphrases Kline [35, p. 19].

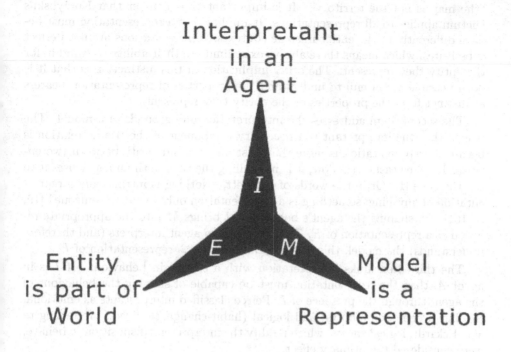

Interpretant
in an
Agent

I

E   M

Entity
is part of
World

Model
is a
Representation

Figure 1.1: Representation as a triadic relation between entity, model and interpretant.

3. His attempt to understand what makes a mental effect an interpretant of some particular representation;

4. His struggle with the problem of interpretation for mental representation;

5. The idea that model and entity are related by two very different sets of relations—semantic relations (such as representing, signifying, referring to, and expressing) and the ground relations in virtue of which those semantic relations hold;

6. His taxonomy of kinds of ground; and

7. His apparent interest in ultimately understanding representation in a completely naturalistic way.

Examining each of these points in turn, the first distinction leads Peirce to consider the character of a model itself. Von Eckardt uses the term "representation bearer" to refer to the properties of the model that belong to the model itself, and not to the entity it represents. In a similar vein, Kline [35] brings attention to the essential difference between a model and an entity by invoking what he calls Korzybski's Dictum, after Alfred Korzybski's [36] warning that "the map is not the territory". It is important to remember that Korzybski's Dictum applies to all representations. It implies that a representation must behave differently to the entity in some contexts. Representations are not perfect substitutes, which means there always exist limits to their ability to stand in for the entity they represent. The other implication of this distinction is that it is both possible and useful to understand the properties of representation bearers as distinct from the properties of the entity they represent.

The second item addresses the interpretation and grounds of a model. The reason that the interpretant is a necessary component of the triadic relation is because a representation is more than just a logical similarity between two entities. If a tree casts a shadow, it is not telling the time until an agent *uses it* to tell the time [1]. Or in the words of Dennett, "Nothing is intrinsically a representation of anything; something is a representation only *for* or *to* someone" [19, p. 101]. By shaping the agent's behaviour, I brings M into the appropriate relation as a representation of E. That is, when an agent interprets (and therefore understands) the model, this grounds the model as a representation of E.

The third item links interpretation with a change in behaviour. If I is in agent A, then the representation must be capable of shaping the behaviour of the agent through the presence of I. Peirce classified interpretants as emotional (feelings), energetic (efforts) and logical (habit-changing) effects. According to Von Eckardt, logical effects, which modify the interpreter's disposition to behave, were considered the primary effect.

The fourth item refers to the problem of infinite regress. While non-mental representation is relatively straightforward, issues arise when M is internal to the agent. The difficult question one faces is "what interprets a mental representation?" If mental representations are interpreted in the same way as non-mental representations, this gives rise to an infinite regress of thoughts interpreting thoughts [63, p. 282].

Separating semantic and ground relations, as noted in point five, allows one to account for how a semantic relation can come to exist. In order for a model M to produce an interpretant I in an agent A, it is necessary for A to understand the representation, which requires A to have knowledge of what the ground is. The following example clarifies this account [63, p. 156]:

> For example, suppose I see a photograph. To understand that photograph I must know (in some sense) that there are both a causal relation and a similarity relation between the photograph and its subject, and I must know (in some sense) the respects in which the photograph is a causal effect of and is similar to its subject. If I know all that, then I will be able to form a belief or a thought about the subject of the photograph (that is, who or what the photograph represents)—specifically, that there was such a subject and that this subject looked a certain way at the time the photograph was taken. In other words, by considering the photograph in conjunction with its ground I come to be in a relation to the object it represents.

With respect to item six, according to Peirce, there exist three kinds of pure ground: iconic, indexical, and symbolic [63, p. 150]. Icons, such as diagrams and images, are models grounded by their intrinsic (first order) similarity to the entity they represent. An index, such as a weathervane, signifies an entity because of a causal or spatiotemporal connection between the index and the entity. Symbolic representations, such as words, are grounded by convention. Symbols act as models only because of the way they are consistently interpreted, which can then generate regular effects on the behaviour of the agent.

In the final item, Von Eckardt interprets Peirce's theory of signs as naturalistic, meaning closely connected to natural science. The naturalistic approach fits neatly with the metaphysical assumption of supervenience outlined in Section 1.2.

I will now propose a definition of representation that reflects Peirce's triadic relation.

Definition 2 (Representation) *A triadic relation between a model, entity and agent. The model substitutes for at least one entity, shaping the behaviour of at least one agent.*

The model *stands in* for an entity, and it always does so *for* an agent, thereby modifying the agent's predisposition to behave. In this definition, the interpretant is implicit in the ability of the model to shape the behaviour of an agent. The model may refer to a class of entities, and may also be shared by multiple agents. However, at least one entity and one agent are necessary for a representation relation.

1.4 External Representation

According to Peirce's triadic relation, the entity E is part of the world, the interpretant I is in the agent, but the location of the model M is unspecified. For the case that M is external to the agent, the triadic relation is relatively straightforward, since the problem of infinite regress does not need to be addressed.

Peirce's typology describes three 'pure' types of grounding relations, which is important for a theory of representation in general. However, in practice, models may incorporate some combination of iconic, indexical and symbolic grounds. The aim of this section is to provide concrete examples of external models, and then show how entities, external models and interpretants have been classified within disparate academic disciplines.

1.4.1 Three kinds of external models

Common models used in representation can be distinguished by implementation rather than pure type. Here, I will assume that a model is somehow simpler than the entity it represents. Although not necessarily true, in practice this is reasonable, since a 1 : 1 mapping in complete detail is in general completely useless (consider a life-sized map of the world, then consider trying to maintain the accuracy of every detail). Even if the representation bearer is not itself simple, practical models must confer some benefit, such as ease of manipulation[3]. Models have deliberate differences and may accentuate salient features, in order to retain only those aspects that are necessary to stand in for the entity. A caricature of a politician and a scale model of an aeroplane are examples of representations that can be understood and manipulated efficiently, which makes them useful substitutes for direct experience under certain conditions.

One special kind of representation is a mathematical model. For example, two contained gas particles can be modelled mathematically by two hard uniform spheres with no internal energy except velocity, in an enclosed continuous four dimensional space (including time). The dynamics of the model constrain its behaviour by conserving momentum and energy, which is transferred along the axis joining the spheres' centres of mass when they collide elastically. The spheres are reflected by collisions with the containing walls. In principle, the model, in conjunction with initial measurements of position and velocity, can be used to predict the outcome of measuring the position and velocity of the particles at any future time. The model is a representation when someone (or more generally an agent) uses the model to stand in for a system of interest. For example, the agent could deduce the value of variables associated with the particles in place of direct observations of the gas particles at future times. In Peirce's typology, mathematical models have symbolic grounds. A mathematical model can always be interpreted as manipulating symbols in a formal system according to syntactic rules[4].

The gas particle dynamics can be represented in at least two other ways. Predictions could also be derived using a physical model, such as two balls on a billiard table. The billiard balls are analog representations, which are not arbitrary and abstract like symbols, but are in some way analogous to their subject. Formally, an analog model must exhibit systematic variation with its task domain [41]. This means the analog model does not have to represent every aspect of the entity in the same units – consider a sun dial, which represents the passage of time as the movement in space of a shadow. Animal testing of pharmaceuticals, architects' scale models and pictures are examples of analog representations, although note that the last two examples can also contain symbolic content, which is usually of a secondary nature. Note that in some analog models, it is possible to view the model from multiple perspectives, while other analog models may fix the perspective in the process of representing an entity. Either way, an agent must use the analog model in place of real world measurements in order to fulfil the representation relation. In Peirce's typology, analog models have iconic grounds.

Another way the gas particles could be represented is using the English lan-

---

[3]An example of a useful 1 : 1 mapping is the conversion between Polar and Cartesian coordinates, which is practical because performing this conversion can often improve the ease of manipulation.

[4]For some areas of mathematics, the corresponding formal system may have an infinite number of axioms, rules or symbols, however these areas of pure mathematics are not practical for forming representations, and for my purposes can safely be ignored.

guage. For the task of predicting particle dynamics, language is quite limited. However, if the particles were at sufficiently low temperature that their movement was frozen, an English description of their configuration could provide a useful representation for an agent. Language can be used to arrange words, which function as labels, to represent objects. Nouns are labels for entities or classes of entities, while the verb phrase of a predicate with two arguments (two nouns) refers to the relationship between the corresponding entities. Labels in isolation can act as signs, which constitute the most primitive form of representation, capable of standing in for only a single idea. Signs are formalised in semiology, whose contemporary form follows much more closely from the work of Saussure [54] than Peirce [29]. When a set of signs is organised into a language with syntactic rules for manipulation and intricate networks of relationships between components, its representational power is qualitatively increased, and is rich enough to be studied in the distinct but related fields of linguistics and structuralist philosophy. An important observation is that the structure of sign systems (languages) does not need to represent the structure of the world. The structure in language is based on the difference between terms, rather than a reflection of structure in the world. This decoupling both provides flexibility of expression within language, while at the same time necessarily limiting its representational nature. Following Saussure, this constructionist view of language is the dominant view in structuralist and post-structuralist philosophy. Note that while mathematics is also a language[5], and both mathematics and language have symbolic grounds, I consider formal systems separately from linguistic representations, because they can play significantly different roles in representation.

Many disciplines have developed explanations of the way external models – mathematical, analog and linguistic – are used by agents to represent their world. The three disciplines I now consider are metaphysics, military theory and systems theory. The terminology and the scope of representations under consideration varies significantly. In spite of this, it is found that Peirce's triadic relation provides a common structure for explaining representation in each case, and also that the external models conform to the three kinds identified in this subsection. Further, the theory of representation in general reveals shortcomings in each of the disciplinary accounts.

1.4.2   Representation in metaphysics

Popper advocated an ontological pluralist doctrine from 1967, which is detailed in *Objective Knowledge* [49] and concisely summarised in [50]. According to Popper, there exists three worlds:

- World 1 is the physical universe, including both living organisms and non-organic matter.

---

[5]This interpretation is made precise in formal language theory.

- World 2 is the world of individual psychology, of mental events, raw feels and thoughts.

- World 3 is the world of abstract products of the human mind, including language, scientific theories, mathematics, paintings and symphonies.

The use of 'world' is indicative of the ontological nature of Popper's distinction. He clearly views each world as consisting of different kinds of stuff, proposing "a view of the universe that recognizes at least three different but interacting sub-universes." [50, p. 143]. The nature of the interactions are causal, and the abstract world is always linked to the physical world via the human mind [50, p. 165]:

> If I am right that the physical world has been changed by the world 3 products *of the human mind*, acting through the intervention *of the human mind* then this means that the worlds 1, 2, and 3, can interact and, therefore, that none of them is causally closed.

Popper contrasts his three world hypothesis with ontological monism (materialism or physicalism) and ontological dualism (mind-body dualism) by saying that the monist only admits world 1, while the dualist only admits worlds 1 and 2. When Popper refers to say a symphony or a sculpture in world 3, this is separate from the world 1 instantiation of the entity. It is only the abstract ideal of the entity that exists in world 3. Thus, world 3 entities are types, which may have many corresponding real world tokens that are imperfect embodiments of their type. The key to Popper's defence of world 3 are the claims that a) abstract entities exist that are not embodied in world 1 or 2, such as the infinite members of the set of natural numbers $\mathbb{N}$; and b) abstract entities have a causal influence on world 1, such as Einstein's equation $e = mc^2$ resulting in the development and use of an atomic bomb.

The three world hypothesis is of interest to us, because it neatly separates the real world entities E that are being represented (world 1), mental interpretations I of those entities (world 2), and external models M that are products of the human mind (world 3), in a way that is compatible[6] with Peirce's triadic relation. However, Popper's cosmology directly contradicts our understanding of physics. In particular, conservation laws and symmetry imply that world 1 is closed, and current theory requires only four fundamental forces (the strong and weak nuclear forces, electromagnetism and gravity) to explain every causal physical interaction. Popper claims that world 3 entities that cannot be embodied in world 1 can nevertheless exert a causal influence on world 1, because they are apprehended by human minds in world 2, which then control causal events back in world 1. But then any physical explanation of a system that includes humans is causally incomplete. Even if one accounted for all of the interactions of the four fundamental forces, there would be a 'residual causality'

---

[6]For both Peirce and Popper, M could be private or shared. However, there is a difference in emphasis: Peirce is mostly concerned with private use, whereas Popper concentrates on shared uses of M.

that remained unaccounted for. This is because abstract entities are not subject to the four fundamental forces, and yet if they have their claimed causal powers, their absence or presence will change the aggregate force acting on bits of world 1 matter. Consequently, one can ask whether it is conceivable that an experiment exists that could test for a residual causality leak from world 1. This would require us to ascertain the presence or absence of an abstract entity, which would require a human mind, without affecting the physical state in the experiment. But in order to say whether the abstract entity was present, the memory would have to be stored in the brain, thus changing the physical state in the experiment (assuming supervenience). In fact, Popper's claim is metaphysical, and unfalsifiable, in contrast to the ideal of scientific conjecture that he advocated. For our purposes, Popper's ontological distinction is stronger than is justified. The same argument applies to the similar, but less sophisticated distinction that Penrose [46] proposes between the physical, Platonic forms, and the human mind.

1.4.3   Representation in military theory

At the other extreme of the academic spectrum, one finds a position that as far as I can ascertain, is advocated only within the relatively isolated discipline of military theory. A central idea in Network Centric Warfare (NCW) [2] and the closely associated, but broader Effects Based Operations (EBO) [58] concepts, is that military actions occur in three domains: the physical, information and cognitive domains. They are based largely on "common sense" and are not rigorously defined. For example, Garstka [25] provides the circular definition: "The information domain is the domain where information lives." This definition is perpetuated in [2]. More sense can be made of Smith's [58, pp. 160-173] interpretation:

> The three domains provide a general framework for tracing what actually goes on in the stimulus and response process inside human minds and human organizations, and how physical actions in one domain get translated into psychological effects and then into a set of decisions in another domain. Understanding this process is important because with it, we can begin to comprehend how people and organizations perceive a stimulus or action and why they respond or react in the way they do and thus, how we might shape behavior.

Smith then defines each domain.

> ... the physical domain encompasses all the physical actions or stimuli that become the agents for the physical and psychological effects we seek to create. ... the actions in the physical domain may be political, economic, and/or military in nature, and all must be equally considered to be objects or events...

> The information domain includes all sensors that monitor physical actions and collect data. It also includes all the means of collating or

contextualizing that data to create an information stream, and all the means of conveying, displaying, and disseminating that information. In essence, the information domain is the means by which a stimulus is recognized and conveyed to a human or to a human organization...

The cognitive domain is the locus of the functions of perceiving, making sense of a situation, assessing alternatives, and deciding on a course of action. This process relies partially on conscious reasoning, the domain of reason, and partially upon sub-conscious mental models, the domain of belief. Both reason and belief are pre-conditioned by culture, education, and experience.

It is clear from these definitions that the physical domain contains the entities E that one needs to represent; the information domain is where external models M are displayed and disseminated; and the cognitive domain is where the models are made sense of – where the interpretants I exist. In both Smith [58] and Alberts et al. [2], the relationship between the domains is seen to be a flow from the physical domain to the cognitive domain via the information domain. Alberts et al. [2, pp. 12-13] establish this flow, and then use it to motivate the central importance of information:

With the exception of direct sensory observation, all of our information about the world comes through and is affected by our interaction with the information domain. And it is through the information domain that we communicate with others (telepathy would be an exception). Consequently, it is increasingly the information domain that must be protected and defended to enable a force to generate combat power in the face of offensive actions taken by an adversary. And, in the all important battle for Information Superiority, the information domain is ground zero.

Disregarding the reference to telepathy and the sales speak, what is Alberts' claim? Direct sensation is claimed to be an exception to the usual flow of understanding from the physical world to the cognitive domain, via the information domain. This description, along with the accompanying diagram – reproduced in Figure 1.2, conjures up visions of the information domain as populated by automated sensors collecting, fusing and disseminating data unaided by human cognition and judgement. Smith is again more cautious, describing the cognitive domain as the locus where data is interpreted and decisions are made, not the information network. However, he maintains the same connections from the physical domain to the information domain, and the information domain to the cognitive domain. This is most explicit in the layered diagrams Smith uses to depict the domains, with the physical domain layer at the bottom, the information domain layer in the middle, and the cognitive domain layer on top.

Interestingly, the three domain model is derived[7] from Fuller's [24] book *The Foundations of the Science of War*, in which Fuller described a trinity between

---

[7]In [25], Garstka notes that "A key element of the model is a focus on three domains: the physical domain, the cognitive domain, and the information domain. This conceptual model

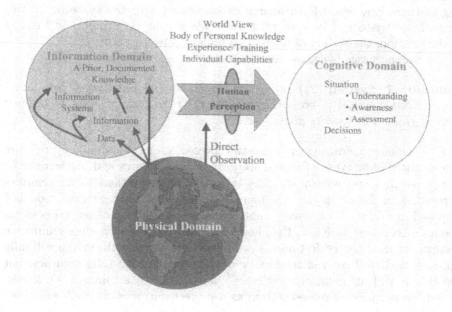

Figure 1.2: The three domains of warfare, after Alberts *et al.* [2].

the three spheres of man. Because they are both based on a triadic relation, the structure of Fuller's three sphere theory of warfare is structurally analogous to Peirce's theory of signs. However, in the modern day interpretation of Fuller, the triadic relation between the domains of warfare has been reduced to two dyadic relations: the physical–information and information–cognitive relations. This intended "refinement" of Fuller's conceptual model has only concealed the essential nature of representation as an irreducible triadic relation.

1.4.4 Representation in systems theory

In his book on multidisciplinary thinking, Kline [35, p. 16] notes three uses of the word 'system' within science, which are relevant to the role of external representations. So that they can be compared, he gives them three separate labels. The first conception, the most common use by scientists outside the systems community, is:

Definition 3 (System (1)) *The object of study, what we want to discuss, define, analyse, think about, write about, and so forth.*

Kline refers to this understanding with the label **'system'**, which for example could refer to the solar system, a communicating vessel, an ecosystem, or an

---

builds upon a construct proposed initially by J.F.C. Fuller in 1917, and refined in *Measuring the Effects of Network-Centric Warfare.*"

operating system. In fact, according to Kline, a **system** can be anything, as long as there is a well defined boundary associated with the **system**. In this thesis, I use 'system of interest' to denote this meaning of system. In Peirce's triadic relation, the system of interest corresponds to the entity E.

The second usage is defined as:

Definition 4 (System (2)) *A picture, equation, mental image, conceptual model, word description, etc., which represents the entity we want to discuss, analyse, think about, write about, etc.*

Kline coins the term 'sysrep' to mean representations of systems. Sysreps are "one of three basic types of representation: words, pictures and mathematics": that is, sysreps are models M. The types of representations Kline identifies correspond to the categories of language, analog and mathematical models I proposed in Section 1.4.1, except that pictures are only one of several possible analogs. According to Kline, the ideal aim of a sysrep is to perfectly mirror a **system**, where "[b]y 'perfect mirror' we mean not only that the sysrep will fully represent each and every characteristic of the **system** with total accuracy, but also that it will represent nothing more" [35, p. 18]. The common – but misguided – conception of representation as a perfect mirror is critically examined in Section 1.5.2.

The third usage is the most general conception, which is consistent with attempts to define the meaning of system within the systems community:

Definition 5 (System (3)) *An integrated entity of heterogeneous parts which acts in a coordinated way.*

Kline uses the label 'system' or 'systemic' for this conception, where a systemic property is an emergent property, which is a property of the whole but not a property of the components of the system.

The final concept Kline invokes is a schemata, which denotes "all the ideas in a person's head which are used to represent and interact with the world" [35, p. 31]. Example schema include words, relational ideas, behavioural routines and medical diagnosis. Kline then answers the question: "What is the relation of a sysrep to schemata in the mind? A sysrep is a particular kind of schemata, a very special class of the totality of the schemata we construct in our minds." Kline defines non-mental representation as a special class and an extension of mental representation. This approach is problematic, because mental representation is actually more difficult to understand than external representation. It makes more sense to explain mental representation in terms of the more straightforward case of external representation, even if mental representation precedes external representation from a chronological view.

In Kline's view, the relation between entities, models and their interpretants is as follows. Scientists view the world as being comprised of **systems**, which are interpreted using mental schemata. Schemata enable complex interactions with the world, but are formed using largely non-conscious mechanisms, and may be fuzzy and unstructured. When we go to the trouble of making a schemata explicit and shared in a structured social environment, it becomes a sysrep (which must still be interpreted by people). The goal of forming sysreps is to mirror the **system**, so that ideally the behaviour of the sysrep and the **system** are identical. While Kline is right to distinguish between systems, system representations, and system interpretations, the details of how they interact are not consistent with Peirce's triadic relation.

Burke [14] has formalised and refined the systems approach to understanding representation, in a clearly articulated conceptual model. He offers the following definitions for entity, system, system description and model [14, pp. 9-12]:

Definition 6 (Entity) *An entity is any object that has existence in the physical, conceptual or socio-cultural domains.*

Definition 7 (System (4)) *A system is an idealisation of an entity as a complex whole.*

Definition 8 (System description) *A system description is a representation of a system.*

Definition 9 (Model) *A model is an idealisation and/or representation of an entity.*

Four implications follow from these definitions. Firstly, because systems are idealisations of entities, they are abstractions that have no physical existence [15]. Systems are not part of the furniture of the world, they only exist inside minds. Stated another way, a system is a way of looking at the world [64]. Secondly, an entity can be idealised as a system in multiple ways: there is no unique systems view for any entity. Thirdly, and most importantly for this discussion on representation, both systems and system descriptions are considered to be models by Burke. The difference is that external models (a system description) presuppose the existence of a corresponding idealisation (a system). This is equivalent to requiring that external models M require an interpretant I in order to represent an entity E. Therefore, Burke's system theoretic interpretation is consistent with Peirce's triadic relation for external representation. Fourthly, Burke defines system descriptions to be derived from systems (idealisations of entities), rather than directly from entities. This implies that the system description can only capture aspects of the entity that have already been captured in the system. Consequently, a system description can be interpreted as a system that has been further abstracted from the entity it represents.

1.4.5 Summary of external representation

I will conclude the discussion of external representation by comparing the distinctions that have been identified above in different disciplines. The most notable commonality is that in each case, exactly three categories have been necessary to explain external representation, and furthermore these categories can be aligned with the entities, models and interpretants of Peirce's triadic relation. Of course, this has more to do with the selective nature of my literature survey than uniformity of approach. Descartes' [21] dualism was unconcerned with external representations, while Rosen's [53] Modeling Relation between the formal systems of science and the natural systems they represent attempted to explain external representations without explicit reference to the human mind or interpretants. Nevertheless, each of the approaches I have covered supposes that things are naturally considered as belonging either to the physical, the mental, or the social products of the mental. The physical world contains the entities that one would like to represent, external models are social products that can be shared, but they must be interpreted by someone or some agent to count as a representation.

There is an important way in which the domains in military theory differ from the accounts of representation by Popper, Kline and Burke. Interactions between the physical domain and the cognitive domain are mediated by the information domain. In contrast, the other accounts explain external models as products of the human mind. Physical entities must be conceptualised before they can be externally represented. Because militaries functionally separate the collection of information from decision-making, the role of human conceptualisation in information collection that mediates between the physical and information collection is easily ignored. But without human intervention and judgement there is only data, not representation or information, and automation can reduce but not eliminate human participation in constructing representations[8]. In view of Peirce's triadic relation, each of the alternative accounts considered aspects of this relation, but none are as comprehensive as Peirce's theory of signs.

External models have been variously held to be: abstract products of the human mind; information bearing artifacts; the socio-cultural environment; a specially precise subset of mental representation; a mirror that reflects part of the world; and a mental representation reduced by additional simplifying assumptions, which is explicit and shared. However, most of these assertions are not entirely accurate. Definitions, such as Kline's, that attempt to explain external representations with respect to mental representation are not enlightening, because the cause is more complicated than the effect. Peirce's typology of iconic, indexical and symbolic pure forms, and my categories of formal, analog and linguistic models, provide a framework for understanding external representation, which is sufficiently general to account for representation across disparate disciplines. Within this framework, an external model is most accurately conceived of as a grounded representation bearer external to the agent who interprets the model. Less formally, an external model is an equation, analog or description that represents something for an agent and thereby modifies its behaviour.

---

[8]This is a point that Polanyi [48, p. 20] makes well, and an example is the automation of the photo-finish for horse races, which still required human judgement in a case where one horse was fractionally in front, but the other extended further past the finish line due to a thick long thread of saliva coming from the horse's mouth. It would seem that such semantic ambiguities cannot be satisfactorily resolved by syntactic processors.

1.5 Internal Representation

Representation plays an important explanatory role in biology. From the perspective of a living agent, the world contains limited essential resources of energy and matter for survival and reproduction, as well as threats to survival such as predators and other harmful energy sources. It is easy to see that the ability to sense qualities of the immediate environment and to control locomotion with context sensitive behaviour confers a significant relative selective advantage. Bacteria that follow a chemical or light gradient can be viewed as performing very basic representation: chemical reactions triggered by the local environment stand for greater expected concentrations of non-local useful energy which cannot be directly detected. An agent that can sense distal features of its environment, using passive or active sensors to detect patterns of incoming energy such as light photons or sound waves, can secure an even greater selective advantage. Whereas proximal sensory information requires an agent to 'bump' into a threat before it can react to it, an agent that can sense a threat at a distance can avoid the threat entirely.

However, distal information is noisy, incomplete and intermittent. Just because a predator becomes occluded by vegetation does not secure the safety of its prey. Current sensory input alone is inadequate for determining the best action in any context. By constructing an internal representation of its environment, an agent can continue to act appropriately in the absence of direct sensory stimuli.

This story of representation in biology is inspired by the accounts of Dennett [20, pp. 177-182] and O'Brien and Opie [42], which suggest that representation is the problem that the brain is intended to solve. There is some empirical support for this conjecture in the form of the sea squirt *Ciona intestinalis*. The tadpole larva has a central nervous system of about 330 cells that controls locomotion. Once it attaches to a permanent object, it undergoes a metamorphosis that has been loosely described as eating its own brain (the cerebral ganglion is broken down and reused), since it no longer needs sensorimotor control, and therefore has no need to represent its environment.

Given this story, one may ask how the brain represents. This question has generated the most sophisticated conversation about internal representations, and has been especially preoccupied with the human brain. The Representational Theory of Mind, or representationalism, dates back at least to Aristotle [47]. The proposed answers of contemporary cognitive science divide into three main camps. They are Good Old Fashioned Artificial Intelligence (GOFAI), also known as symbolic, classical, or conventional cognitive science; connectionism; and the dynamical systems hypothesis.

When the representation is internal to the agent, one is faced with the question of what interprets the model. If internal models are interpreted in the same way as external models, then this leads to infinite regress, because the interpretant is also an internal model that requires its own interpretant, and so on. Von Eckardt [63, p. 283] describes two alternative resolutions to the infinite regress problem Peirce considered, and relates these to analogous moves in

247

contemporary cognitive science.

The first solution is to weaken the definition of interpretant, to be a *potential* rather than an *actual* interpretant. The regress still consists of an infinite series of representations, but it is now easier to reconcile the associated interpretants, because they do not need to actually exist. This solution is reiterated by Cummins in cognitive science.

The second solution is "to find something that can function as an interpretant but which is not, itself, also representational and therefore in need of interpretation" [63, p. 283]. Peirce suggests that the only candidate for this is a habit-change. Specifically, Von Eckardt argues it must be a modification in the tendency to act in ways dependent on the content of the representation. The habit-change does not need to affect external behaviour; changes to mental habits (processes that generate other internal representations) also count. However, in order to eventually curtail the regress, internal models must ultimately be interpreted by shaping the agent's external behaviour. A very similar solution is suggested by Dennett, which Von Eckardt claims is the widely endorsed solution in cognitive science. Further, Von Eckardt [63, p. 290-302] shows in detail how this solution can handle the regress problem. Briefly, this involves demonstrating that:

- Interpretant I of model M is producible by M; and

- I is related to both the agent A and M, such that by means of I the content of M can make a difference to the internal states or the external behaviour of A towards the entity E.

Von Eckardt establishes this is the case for both conventional (symbolic) and connectionist machines. I will now provide a short introduction to GOFAI and connectionism, the two strongest advocates of representationalism.

1.5.1 Representationalism

How can a particular state or event in the brain represent one feature of the world rather than another? And whatever it is that makes some feature of the brain represent what it represents, how does it come to represent what it represents?

Daniel Dennett

These are the questions of representationalism, a position that assumes that the mind is a representational device, and that the brain has representational content. They are exceptionally difficult questions, because the mechanisms behind brain functions such as learning, memory and computation in the brain are currently poorly understood. Consequently, the mechanisms underlying representation are equally opaque. Also, under almost any metric, the human brain

248

rates as the one of the most complex entities studied in science[9]. For a deeper discussion of representationalism than I can afford here, see Cummins [17].

As is the case for most enduring themes of Western philosophy, the first records of representational theories of mind are found in the writings of Aristotle [4]. In Book III, part 4, Aristotle describes the part of the human soul that thinks and judges: νοῦς or the mind. According to Aristotle, the mind is "capable of receiving the form of an object; that is, must be potentially identical in character with its object without being the object." This statement clearly demonstrates Aristotle's use of the distinction between a model and the entity it represents. By form, Aristotle refers to the properties of the object, as opposed to its material substance. In Aristotle's metaphysics, the immaterial mind knows something when it takes on the form of that object, such that it represents the object in virtue of their similarity, in exactly the same way that a picture can represent a scene (Peirce's iconic grounds). "To the thinking soul images serve as if they were contents of perception ... That is why the soul never thinks without an image." Berkeley [9] and Hume [33, 34] both extended this Aristotelian conception to argue that all mental contents are images in the mind, and that they are representations in virtue of their resemblance to perception. The inherent weakness of basing mental content on similarity can be seen by probing the mechanisms that could imbue mental images with the same properties as the objects they represent. Images presented to an immaterial mind are not so much an explanation as a metaphor, where thinking is like putting on a theatre for the Eye of the Mind.

In contrast, Hobbes [27, Chapter V] and Leibniz [37] advanced the idea that everything done by the mind is a computation. In this view, thought proceeded by symbolic manipulations analogous to the additions and subtractions of the new calculating devices – in modern parlance the mind was seen as an "automatic formal system" [26]. Notably, this reframed the question of representationalism to propose a mechanical and material explanation of mental processes. This provided a crucial step towards a science of cognition, because it opens up the possibility that certain features of cognition could be reproduced artificially.

The link between computation and representation is important but subtle. Because of the universality of Turing's conceptual model of digital (symbolic) computation, it is a common assumption that all computation is equivalent to a Universal Turing Machine. However, as O'Brien and Opie [43] correctly point out, this does not account for analog computation. They propose a definition of computation in general, which is broad enough to capture both analog and digital computation, but still sufficiently constrained to differentiate computation from the vast majority of physical systems – intestines, microwave ovens, cups of tea, etc. – that are not involved in computation.

> [T]here are two distinctive features of computational processes (as opposed to causal processes in general). First, they are associated with representing vehicles of some kind. Second, and more import-

---

[9]See Bar-Yam [7] for estimates of the complexity of the brain compared with other systems.

antly, computational processes are shaped by the contents of the very representations they implicate. We thus arrive at the following characterisation:

Computations are causal processes that implicate one or more representing vehicles, such that their trajectory is shaped by the representational contents of those vehicles.

This characterisation of computation makes explicit the link between computation and representation. Computations are those processes involving representing vehicles (models), such that the outcome of the process depends on the content of the model. Representation is inherent in computational processes, and computation is the mechanism that causally links the contents of models to changes in the behaviour of the agent that interprets the model. Representation and computation are a package deal: a commitment to a computational theory of mind entails a commitment to representationalism.

Although conceived in the 17th century, it was not until the mid 20th century that the computational idea rose to prominence. The initial hype associated with the AI movement had a profound impact on 20th century cognitive science, such that computational theories of mind were predominantly based on algorithmic symbol manipulation. The Universal Turing Machine [61] provided a theoretical basis for universal symbol-based simulators of human intelligence, while exponential increases in computing power dramatically expanded the application of computer algorithms towards focussed engineering tasks that had previously required the application of the human mind.

Yet simulations that could be confused with intelligent humans have not materialised. AI researchers began to hit some fundamental walls: general intelligence appeared to require fast, situated, unencapsulated reasoning, where automated formal systems were slow, abstract, and only capable of manipulating the initial axioms they were given according to fixed rules. Coinciding with a growing dissatisfaction with the ability of the products of AI to live up to expectations, several alternatives have been advanced within cognitive science, and symbolic computational models of cognition began to be referred to as GOFAI.

Connectionism, or Parallel Distributed Processing (see for example [39]), which is based on highly abstract networks of artificial neurons, presents an alternative paradigm for modelling cognition, which can be interpreted as performing analog computation. Connectionist models are inspired by current understanding of the architecture of the brain, and are described by Dennett [20, p. 269] as blazing the first remotely plausible trails of unification between the mind sciences and the brain sciences. Different kinds of connectionist networks have been shown to have content addressable memory [30]; provide universal function approximation [31]; degrade gradually when damaged; and distribute representations across the set of connection weights, which decouples representations from individual symbols. Due to their parallel processing, connectionist networks are also very fast. The theoretical model of connectionism, an artificial neural network with real connection weights, has been proven to be capable of hypercomputation [57] – that is, able to compute functions that Turing machines cannot. Of course, such machines are not practical, since real numbers in general require infinite information, and there are also a number of issues artificial neural network implementations suffer from. They are almost always simulated on a digital computer, which implies these instantiations are equivalent to Turing machines; they learn reliably only under supervision; and they are usually treated as black boxes, because their behaviour is not currently well understood. Of course, there are philosophical concerns as well. For example, Fodor and Pylyshyn [23] criticise connectionism because it cannot explain systematicity: the feature of human cognition whereby the ability to think one thought entails the ability to think of numerous logically related thoughts, such as its converse.

The most recent alternative to both GOFAI and connectionism is the dynamical systems hypothesis [62]. However, advocates of the dynamical systems account are often explicitly critical of explanations involving representation, so a discussion of dynamical systems is deferred to Section 1.5.2 on anti-representationalism.

In summary, representational theories of mind have been proposed that are based on symbolic manipulation and analog covariance. GOFAI and connectionism agree that mental contents can stand in for, and stand in relation to real world objects. They also assume that psychological processes are computations that represent aspects of the external world.

1.5.2 Anti-representationalism

> There is no harm in saying of good tools and good moves that they are also good representations, but nothing interesting is conveyed by this choice of idiom, and its employment should not tempt us to construct theories about how representation works.
>
> Richard Rorty

Accounts of representation in cognitive science and artificial intelligence have been criticised as a basis for biological behaviour on a number of fronts. Brooks [12] summarises one key idea against representation in the physical grounding hypothesis: the world is its own best model, the trick is to sense it appropriately and often enough. The first part is true but uninteresting, because it is the trivial case where the representation relation degenerates into a diadic relation between A and $E \equiv M$. The second part is important, because it emphasises the need for agent decisions to be grounded. However, the physical grounding hypothesis does not and can not dispose of representation entirely. Even if the world is used as its own model, the agent needs to interpret the meaning of its observations. Constructionist accounts of vision (see for example [45]) argue that the process of perception involves significant construction by the observer. In their critique of a simplistic but common conception of "pure vision" – essentially the idea that the visual system is a bottom-up hierarchy designed to fully mirror the visual scene – Churchland *et al.* [16] provide an alternative account that they label interactive vision. Some of the constructive characteristics of interactive vision are: visual fields are highly non-uniform; vision is exploratory and predictive; the motor system and the visual system are entangled; sensory processing is more like a recurrent network than a hierarchy; and vision cannot be neatly separated from other brain functions. Consequently, the process of sensing the world appropriately is in fact one of the major sources of representational activity in the brain [44]. Also, in Section 1.3 I gave a number of reasons why internal representations can be convenient, even if they are not perfect substitutes for unmediated access to the real world.

Van Gelder [62] denies that cognition involves computation or representation, advancing an alternative dynamical systems hypothesis. In this account, rather than interpreting cognitive states as symbols, they are treated as quantifiable states of a nonlinear dynamical system. Van Gelder uses the Watt governor, depicted in Figure 1.3, to illustrate his thesis. The Watt governor is a mechanical device that maintains a constant speed for a flywheel despite fluctuations in both steam pressure from the boilers and the engine workload. The Watt governor was a pivotal invention during the industrial revolution that allowed the generation of reliable, smooth and uniform power. The Watt governor works because a spindle is geared into the flywheel such that the spindle rotates proportionally to the speed of the flywheel. The faster the spindle rotates, the more centrifugal force it generates, raising the spindle of the flywheel. Because the spindle is directly linked to the throttle valve, the faster the spindle rotates, the higher its arms rise, the more the valve is closed, restricting the flow of steam. As the speed of the flywheel decreases, so too does the spindle, the arms fall, opening the valve and increasing the flow of steam. Thus, a steady state for the speed of the flywheel exists and the Watt governor maintains the steady state by exerting negative feedback on any deviation from the steady state.

Van Gelder compares this mechanical device, which he classifies as a dynamical system, with a hypothetical computational device capable of performing the same function. The computational device would follow an algorithm that depends on representation.

> The very first thing it does is measure its environment (the engine) to obtain a symbolic representation of current engine speed. It then performs a series of operations on this and other representations, resulting in an output representation, a symbolic specification of the alteration to be made in the throttle valve; this representation then causes the valve adjusting mechanism to make the corresponding change [62, p. 350].

In contrast, the mechanical device is non-representational. Van Gelder gives four reasons: representation is not needed to fully account for the operation of the Watt governor; the obvious correlation between arm angle and engine speed is not representational because representation is more than mere correlation; the simple correlation only obtains in the steady state; and the arm angle cannot

252

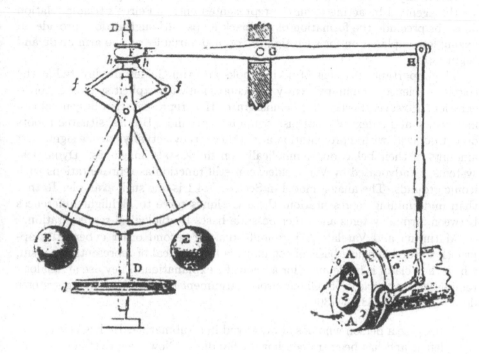

Figure 1.3: The Watt centrifugal governor for controlling the speed of a steam engine, after [22] as reproduced in [62].

represent engine speed because the two quantities are coupled.

Of these, the first three reasons are not persuasive. Just because an explanation of the Watt governor within some frameworks do not need the concept of representation does not imply that representation *cannot* be used to explain the same process. After all, none of the compound objects – such as spindles and throttles – are necessary concepts in the quantum mechanical wavefunction of a Watt governor. The second point does nothing to disprove representation occurs, it merely demands a higher standard of proof than demonstrating correlation, while the third point only notes that any correlation is not simple.

The forth reason is the most interesting. Van Gelder observes that "the angle of the arms is at all times determining the amount of steam entering the piston, and hence at all times both determined by, and determining, each other's behaviour." Because of this circular causality, Van Gelder claims that representation is "the wrong sort of conceptual tool to apply". When representation is thought of as a mirror, it does indeed seem wrong for the mirror to determine any part of the mirrored entity, because there is an asymmetry in their relationship. However, under the conception of representation as a triadic relation, it is *necessary* for the model to change the behaviour of an agent, and *possible* for the agent to be acting upon the represented entity. Peirce's triadic relation does not preclude the formation of feedback loops, although it does provide an incomplete explanation for such tightly coupled variables as the arm angle and engine speed.

The important criticisms of both Brooks and van Gelder are directed at the cognitive science community's early preoccupation with explicit symbolic representation. However, Section 1.3 demonstrates that representation in general can have iconic and indexical – not just symbolic – grounds. Brooks' situated robots do not do away with representation altogether – they actually encode significant amounts of their behaviour symbolically on finite state machines. Dynamical systems, as advocated by Van Gelder, can still function as representations with iconic grounds. The analog model in Section 1.4.1 is one such example. Rather than undermining representation, these critiques serve to highlight differences between formal systems and other possible bases for biological representation.

Maturana and Varela's [38] ground-breaking second order cybernetics approach to the biological basis of cognition is also critical of representationalism, which they claim is inadequate for a scientific explanation. They use an analogy reminiscent of Searle's [55] Chinese room argument to claim that living systems do not represent [38, p. 136]:

> Imagine a person who has always lived in a submarine. He has never left it and has been trained how to handle it. Now, we are standing on the shore and see the submarine gracefully surfacing. We then get on the radio and tell the navigator inside: 'Congratulations! You avoided the reefs and surfaced beautifully. You really know how to handle a submarine.' The navigator in the submarine, however, is perplexed: 'What's this about reefs and surfacing? All I did was push some levers and turn knobs and make certain relationships between indicators as I operated the levers and knobs. It was all done in a prescribed sequence which I'm used to. I didn't do any special maneuver, and on top of that, you talk to me about a submarine. You must be kidding!'

This analogy works by specifying an overly narrow system boundary. The adequacy of the navigator in avoiding the reefs cannot be explained unless the boundary is expanded to include the process that generated the prescribed sequence of knob turns and lever pushes. Specifically, the person in this example only becomes a navigator once they have been *trained*. But then it is easy to see that the precise purpose of training the navigator in the sequence of actions is *to stand in for* observations of the reefs and the depth below sea level, and thereby modify the submarine's behaviour.

A more serious threat to representationalism is anti-representationalism, which has been advocated by Davidson, and even more forcefully by Rorty [51].

Anti-representationalism holds that any statement about the world is an inseparable cohabitation of subject and object, rather than correspondence between an object and a model. Rorty rejects the 'mirror' metaphor of knowledge, where knowledge is a reflection of the mind-external world. According to Rorty this metaphor, which we have already seen used explicitly by Kline above, is the central metaphor for representationalism. Rorty criticises what he calls the Aristotle-Locke analogy of knowledge to perception,

> ...the original dominating metaphor as being that of having our beliefs determined by being brought face-to-face with the object of the belief (the geometrical figure which proves the theorem, for example). The next stage is to think that to understand how to know better is to understand how to improve the activity of a quasi-visual faculty, the Mirror of Nature, and thus to think of knowledge as an assemblage of accurate representations. Then comes the idea that the way to have accurate representations is to find, within the Mirror, a special privileged class of representations so compelling that their accuracy cannot be doubted. These privileged foundations will be the foundations of knowledge, and the discipline which directs us toward them—the theory of knowledge—will be the foundation of culture. The theory of knowledge will be the search for that which compels the mind to belief as soon as it is unveiled. Philosophy-as-epistemology will be the search for the immutable structures within which knowledge, life, and culture must be contained—structures set by the privileged representations which it studies. The neo-Kantian consensus thus appears as the end-product of an original wish to substitute *confrontation* for *conversation* as the determinant of our belief [51, p. 163].

It should be noted that Rorty's attacks are not directly focussed on the cognitive science debate on mental contents, which the previous critiques have participated in. Rorty, who is trained in analytic philosophy, is more concerned with structuralist and linguistic attempts to ground knowledge as representation. For example, Rorty [52] cites Brandom's [11] characterisation of the representationalist school as saying that "the essential feature of language is its capacity to represent the way things are." Proponents of this school are taken to be Frege, Russell, Tarski and Carnap, who are contrasted with Dewey, Wittgenstein and Sellars, who view language as a set of social practises. However, it was noted in Section 1.4.1 that there is no necessity for language to be representational. Further, Rorty's target is much larger than just the philosophy of mind – he seeks to question the legitimacy of transcendental (Kantian) epistemology as distinct from psychology, and advocates that the only constraints on knowledge are essentially conversational in nature. Rorty rejects the very idea of a theory of knowledge, truth or rationality.

Is it possible to salvage representationalism from these attacks? Rorty seeks an *a priori* defeat of representation, but there is an empirical component to the question of whether minds represent. Languages are not reflections of reality, in the sense that they are not mirrors that can be polished to provide a True representation of the world. However, it does not immediately follow that formal systems, analogs or minds can not represent in any meaningful way.

A stronger defence of representationalism can be made by carefully articulating what work the representation relation needs to perform. From the perspective of an agent faced with a decision, consider two different processes for choosing between the alternatives. For concreteness, suppose the agent is a frog near the edge of a cliff choosing whether to jump forward to the left or right. One choice is safe but the other choice will result in certain death. Using the first process, suppose that the frog, like one of Brooks' situated robots, is able to make its decision by using the world "as its own best model". Brooks' idea of sensing the world often enough is similar to Van Gelder's claim that dynamical coupling and feedback can do the same job as representation, without requiring extensive planning or computation. In the case of the frog, it is a pretty simple and efficient process: both alternatives are sensed and the apparently less perilous alternative is immediately acted upon. The process is memoryless[10], so it can be repeated every time the frog lands. It can work when several conditions are met: if the frog can sense at least as far as it leaps; if the sensory comparison is reliable; and if the environment is sufficiently stationary while the frog is airborne.

If the latter condition is violated, nothing can guarantee the safety of the frog if it continues to leap. However, by using an alternative representational process, the frog may stand a chance even when the first two conditions do not hold. Suppose that it is a dark and foggy night, so that the frog cannot reliably sense the relative merits of jumping left and right. Fortunately, the frog has taken this path many times before, and remembers the sequence of left and right jumps that have got it home safely in the past. Then recognition of the starting location and recollection of this sequence can substitute in the decision-making process for sensing the alternatives at every step. The sequence can act as a model which when interpreted by the frog can stand in for the currently unreliable sensory information. In this process, representation must still be grounded by having previously sensed the alternatives using the first process. However, it can no longer be memoryless. By maintaining an internal model, the frog amplifies the applicability of sensory information beyond immediate local sense-response reflexes, to affect behaviour non-locally in space or time. Representation allows the agent to do more with less sensory information, to fill in gaps and to generalise new information. In other words, the real work that representation is doing is inference.

Rorty is right to reject the metaphor of representation as a mirror, reflecting the nature of reality for Descartes' Eye of the Mind. But representation is not a mirror, its purpose is not to *reflect* but to *infer*. In this section, I have argued that

---

[10]I should clarify that I mean memoryless in the mathematical (Markovian) sense: the probability of future states only depends on the current state. This is significantly more abstract and general than the meaning of memory in cognitive science.

perception is to some extent constructed, which involves representational activity. I have shown that representation can have non-symbolic grounds, meaning dynamical systems can be a basis for representation. I have examined Maturana and Varela's argument against representation, which relies on an overly narrow definition of system boundary. Finally, I have argued that Rorty's attack on representation is largely directed towards language as representational, and to representation as mirroring. These critiques have merit, but do not challenge the general theory of representation provided by Peirce's triadic relation as an explanation of internal representation.

1.6 Agents

So far I have not specified what I mean by an agent. However, the preceding discussion on representation offers a precise way of characterising agency. Because of the ubiquity of agents in complex systems, this section contains the most important implications of this chapter.

As Peirce has argued, representations require an interpretant, and therefore an agent to perform the interpretation. Thus, there is a sense in which representations (and computations) are relative to a subject – that is, they are subjective [56, p. 92]. But there is an equally objective way that a subject plus a model either does, or does not, represent. By redrawing the system boundary to include the model's user, representation is an intrinsic feature of this system of interest. This is the importance of the triadic relation: because it is irreducible, the system boundary must always extend to include the agent in order to understand representation.

This is why, unless one can identify who is using the model and how it shapes their behaviour, the model cannot be considered to be a representation. Peirce, Von Eckardt, Popper, Smith, Kline and Burke all identify the 'who' with a human mind, which is more generally the case in the literature. I have generalised this to say that a model is always a model employed by an agent. An agent can be a person, but it can also be a group of people, an animal, a cell, a certain kind of robot or a certain kind of physical process (after all, each member of this list is a physical process). The ability to form and use representations appears to be the principle difference[11] between an object and an agent – it is the difference between kicking a rock and kicking a cat (not that either experiment is condoned, except as a thought experiment).

More formally, if an entity's response to a stimulus is directly determined by its current state, and the current state does not include any models, then the entity is not an agent. If stimulus and response are indirectly related because they are mediated by representation, then the entity is an agent.

---

[11]For example, Aristotle [4] says "The soul of animals is characterized by two faculties, (a) the faculty of discrimination which is the work of thought and sense, and (b) the faculty of originating local movement." In my account, representations both encode distinctions and shape movement or behaviour.

Definition 10 (Agent) *An entity that constructs and uses representations to shape its own goal-directed behaviour.*

More will be said about goal-directed behaviour below, but for now note that goal-directed behaviour does not imply that agents only have a single goal: it is merely intended to distinguish between directed and undirected behaviour. It seems there is a continuum, such that entities may have a degree of agency, depending on how indirect the relationship between stimulus and response is, and how sophisticated the representations can become, which is often called the plasticity of the representing medium. I am not overly concerned about the precise demarcation between agent and non-agent. The definition is more useful for comparative purposes, in order to investigate if the degree of agency has increased, and to say that a human has more agency than a cockroach, which has more agency than a virus, which has more agency than the robot Cog [13], which has more agency than a cyclone[12].

The degree of autonomy of an agent refers to freedom of choice or variety, which is made precise by the notion of source coding in information theory. The degree of autonomy is evident in the sensitivity of changes in the behaviour of the agent to changes in its representations. For example, if an agent's model is replaced by any other model (such as its inverse) and yet this has no causal influence on the behaviour of the agent, then the model does not contribute to the autonomy of the agent. If this holds for all models, then the agent is not autonomous. A model must shape behaviour to be a representation and provide the agent with autonomy. In contrast, if any arbitrary desired feasible state within the agent's environment can be achieved by changing only the agent's representations, which then realise the desired state by modifying the agent's behaviour, the agent has maximum autonomy. When autonomy is shared between two or more agents, this is the subject of game theory, and the degree of autonomy of a player is the number of available strategies[13], and any mixed strategy on this set constitutes a model.

The autonomy of agents can lead to philosophical debate about free will and teleology. In view of Hume's compatibilism [33, 34], the autonomy of an agent does not imply the agent is necessarily nondeterministic – that with *exactly* the same internal and external states, two distinct responses to the same stimulus are possible. Instead, a weaker condition holds, namely given different representations, an agent is capable of choosing different actions. To confuse the matter, often it is useful to explain the behaviour of a system as an autonomous agent, even when it is clearly not purposive. For example, Dawkins [18] describes genes as selfish molecules, as if they have minds, which is a form of teleonomic

---

[12]Most people do not consider a cyclone to be an agent, even though it is a self-maintaining, non-equilibrium entity with unpredictable behaviour. The anti-cyclonic Great Red Spot on Jupiter is a structure with a diameter significantly greater than Earth, which has persisted since it was observed by Cassini in 1665. I believe the main reason cyclones are not considered agents is because it is not possible to sustain an interpretation of either goal-directed behaviour or representation for a cyclone and although unpredictable, they are not autonomous.

[13]This game theoretic interpretation assumes the set of strategies is countable, where only strategies that affect the value of the game are counted.

explanation. For my purposes, I will always assume that agency entails a degree of autonomy in Hume's sense, and also implies that the agent is capable of exhibiting goal-directed behaviour.

The notion of goal-directed behaviour has been formalised by Sommerhoff [59], who observed that the essence of goal-directed activity is:

> that the occurrence of the goal-event G is *invariant* in respect of certain initial state variables ($\mathbf{u}_0$) *despite* the fact that G depends on action factors and environment factors that are *not* invariant in respect of $\mathbf{u}_0$. The invariance of G being due to the fact that the transitional effects of changes in $\mathbf{u}_0$ mutually compensate, so to speak.

Sommerhoff realised it was possible to treat goal-directedness as an objective property, independently of the subjective notion of purposiveness of interest to the psychologist. He established three necessary and sufficient criteria for goal-directed behaviour. Firstly, for at least one variable $\mathbf{a}$ associated with the action, and one variable $\mathbf{e}$ associated with the environment, for at least one time t_k,

$$F(\mathbf{a}_k, \mathbf{e}_k) = 0. \tag{1.1}$$

This ensures that the action is capable of compensating for environmental variability. Secondly, $\mathbf{a}$ and $\mathbf{e}$ must be mutually orthogonal, meaning that the value of one of the variables does not determine the value of the other for the same instant. This allows the mechanism for goal-directed behaviour to realise Equation (1.1) for a range of initial conditions. And thirdly, there must be a set S_0 of initial environmental conditions (where $|S_0| \geq 2$), such that each initial condition requires a unique action $\mathbf{a}_k$ which satisfies Equation (1.1). This criterion ensures that the goal could have been achieved from an ensemble of initial conditions, rather than only from the actual initial conditions. $|S_0|$ provides a measure of the degree of goal-directed behaviour: the greater $|S_0|$ is, the more environmental variety the agent can destroy and still achieve its goal. Thus, goal-directed behaviour is underpinned by Ashby's [5, 6] law of requisite variety.

From a stimulus-response perspective, an agent can be thought of as sensing stimuli and acting to produce an appropriate response. The function that maps from sensory inputs and models to output actions is its decision map. The sense, decide and act functions of agents are roughly analogous to detectors, rules and effectors in Holland's [28] complex adaptive systems terminology; perceptual, cognitive and motor components in cognitive science; input (actuating signal), control unit (dynamic element), and output (controlled variable) in control theory; state, policy and action in reinforcement learning [60]; stimulus, organism and response in Hull's [32] version of behavioral psychology; state, mixed strategy and move in game theory; and observe, orient/decide and act (OODA) in Boyd's [10] decision cycle.

The sense, decide, act trinity is a pervasive characterisation of agency that can be related back to Peirce's triadic relation. Without sensing, there is no way to ground representations. Without acting, representation cannot shape external behaviour. Without deciding, representations cannot be interpreted, the agent cannot be autonomous, and its behaviour is not goal-directed. A necessary and sufficient condition for agency is the possession of sense, decide, and act functions. But this is exactly equivalent to requiring that an agent be able to construct and use representations to shape goal-directed behaviour.

In summary, representation and agency have been co-defined. Representations always involve an agent, and agents always represent their environment. The triadic nature of the representation relation is the reason that these definitions cannot be separated. Due to this intimate relationship, a theory of representation is essential to an understanding of the behaviour of agents in a complex system.

Bibliography

[1] ABBOTT, R., "If a tree casts a shadow is it telling the time?", *Unconventional Computation, UC2006* (York, UK,) (C. S. CALUDE, M. J. DINNEEN, G. PAUN, G. ROZENBERG, AND S. STEPNEY eds.), Springer (2006).

[2] ALBERTS, D. S., J. J. GARSTKA, R. E. HAYES, and D. A. SIGNORI, *Understanding Information Age Warfare*, CCRP Publication Series (2001).

[3] ANDERSON, P. W., "More is Different", *Science* **177**, 4047 (1972), 393–396.

[4] ARISTOTLE, *De Anima*, transl. J. A. Smith, eBooks@Adelaide, originally published 350BC Adelaide, Australia (2004).

[5] ASHBY, W. R., *An Introduction to Cybernetics*, Chapman & Hall London, UK (1956).

[6] ASHBY, W. R., "Requisite variety and its implications for the control of complex systems", *Cybernetica* **1** (1958), 83–99.

[7] BAR-YAM, Y., *Dynamics of Complex Systems*, Westview Press Boulder, Colorado (1997).

[8] BERGSON, H., *Creative Evolution*, transl. A. Mitchell, Macmillan London (1911).

[9] BERKELEY, G., *A Treatise concerning the Principles of Human Knowledge*, eBooks@Adelaide, originally published 1710 Adelaide, Australia (2006).

[10] BOYD, COL J. A., "A Discourse on Winning and Losing" (1987).

[11] BRANDOM, R., "Truth and Assertability", *Journal of Philosophy* **73** (1976), 137.

[12] BROOKS, R. A., "Elephants Don't Play Chess", *Robotics and Autonomous Systems* **6** (1990), 3–15.

[13] BROOKS, R. A., C. BREAZEAL, M. MARJANOVIC, B. SCASSELLATI, and M. M. WILLIAMSON, "The Cog Project: Bulding a Humaniod Robot", *Computation for Metaphors, Analogy and Agents*, (C. L. NEHANIV ed.). Springer Berlin, Germany (1999).

[14] BURKE, M., "Robustness, Resilience and Adaptability: Implications for National Security, Safety and Stability (Draft)", *Tech. Rep. no.*, DSTO, (2006).

[15] CHECKLAND, P., *Systems thinking, systems practice*, John Wiley and Sons Chichester UK (1981).

[16] CHURCHLAND, P. S., Ramachandran V. S., and T. J. SEJNOWSKI, "A Critique of Pure Vision", *Large-Scale Neuronal Theories of the Brain*, (C. KOCH AND J. L. DAVIS eds.). A Bradford Book, The MIT Press Cambridge, USA (1994).

[17] CUMMINS, R., *Meaning and Mental Representation*, The MIT Press Cambridge, USA (1989).

[18] DAWKINS, R., *The Selfish Gene*, Oxford University Press Oxford, UK (1976).

[19] DENNETT, D. C., *Brainstorms: Philosophical Essays on Mind and Psychology*, MIT Press Cambridge, USA (1978).

[20] DENNETT, D. C., *Consciousness explained*, Little Brown and Co. Boston, USA (1991).

[21] DESCARTES, R., *A Discourse on Method: Meditations and Principles*, transl. J. Veitch, J. M. Dent and Sons London (1912).

[22] FAREY, J., *A Treatise on the Steam Engine: Historical, Practical, and Descriptive*, Longman, Rees, Orme, Brown, and Green London, UK (1827).

[23] FODOR, J. A., and Z. PYLYSHYN, "Connectionism and Cognitive Architecture: A Critical Analysis", *Cognition* **28** (1988), 3–71.

[24] FULLER, J. F. C., *The Foundations of the Science of War*, Hutchinson and Co. Ltd. London (1925).

[25] GARSTKA, J. J., "Network Centric Warfare: An Overview of Emerging Theory", *Phalanx* **33**, 4 (2000), 1–33.

[26] HAUGELAND, J., *Artificial Intelligence: The Very Idea*, MIT Press Cambridge, USA (1985).

[27] HOBBES, T., *Leviathan*, University of Oregon Oregon, USA (1651).

[28] HOLLAND, J. H., *Hidden Order: How Adaptation Builds Complexity*, Helix Books, AddisonWesley Reading, USA (1995).

[29] HOOPES, J. ed., *Peirce on Signs: Writings on Semiotic by Charles Sanders Peirce*, University of North Carolina Press Chapel Hill, USA (1991).

[30] HOPFIELD, J. J., "Neural networks and physical systems with emergent collective computational abilities", *Proceedings of the National Academy of Sciences of the USA* **79**, 8 (1982), 2554–2558.

[31] HORNIK, K., M. STINCHCOMBE, and H. WHITE, "Multilayer feedforward networks are universal approximators", *Neural Networks* **2**, 5 (1989), 359–366, 70408.

[32] HULL, C., *Principles of Behavior*, Appleton-Century-Crofts New York, USA (1943).

[33] HUME, D., *A Treatise of Human Nature*, Reprint Oxford University Press, 1978 Oxford, UK (1741).

[34] HUME, D., *An Enquiry Concerning Human Understanding*, Ed. P. H. Niditch, Reprint Clarendon Press, 1975 Oxford, UK (1777).

[35] KLINE, S. J., *Conceptual Foundations for Multidisciplinary Thinking*, Stanford University Press California, USA (1995).

[36] KORZIBSKI, A., *Science and Sanity*, The International Non-Aristotelian Library (1948).

[37] LEIBNIZ, G. W., *Dissertatio de Arte Combinatoria*, Leipzig, Germany (1666).

[38] MATURANA, H., and F. VARELA, *The Tree of Knowledge: The Biological Roots of Human Understanding*, transl. J. Young, Shambhala Publications, originally published 1988 Boston, USA (1984).

[39] MCCLELLAND, J., and D. RUMELHART eds., *Parallel Distributed Processing: Explorations in the Microstructures of Cognition*, MIT Press Cambridge, USA (1986).

[40] MCLAUGHLIN, B., and K. BENNETT, "Supervenience", *The Stanford Encyclopedia of Philosophy*, (E. ZALTA ed.). The Metaphysics Research Lab (2006).

[41] O'BRIEN, G., "Connectionism, Analogicity and Mental Content", *Acta Analytica* **22** (1998), 111–131.

[42] O'BRIEN, G., and J. OPIE, "Notes Toward a Structuralist Theory of Mental Representation", *Representation in Mind: New Approaches to Mental Representation*, (H. CLAPIN, P. STAINES, AND P. SLEZAK eds.). Elsevier Science (2004).

[43] O'BRIEN, G., and J. OPIE, "How do connectionist networks compute?", *Cognitive Processing* **7**, 1 (2006), 30–41.

[44] OPIE, J., "Personal communication" (2006).

[45] PALMER, S. E., *Vision Science: Photons to Phenomenology*, MIT Press Cambridge, USA (1999).

[46] PENROSE, R., *Shadows of the Mind*, Oxford University Press Oxford, UK (1994).

[47] PITT, D., "Mental Representation", *The Stanford Encyclopedia of Philosophy* (E. ZALTA ed.), (2005).

[48] POLANYI, M., *Personal Knowledge: Towards a Post-Critical Philosophy*, Routledge London, UK (1962).

[49] POPPER, K. R., *Objective Knowledge*, Clarendon Press Oxford, UK (1972).

[50] POPPER, K. R., "Three Worlds", *The Tanner Lecture on Human Values* (The University of Michigan,), (1978).

[51] RORTY, R., *Philosophy and the mirror of nature*, Princeton University Press Princeton (1979).

[52] RORTY, R., "Representation, social practise, and truth", *Objectivity, Relativism and Truth: Philosophical Papers Volume 1*. Cambridge University Press Cambridge, UK (1991).

[53] ROSEN, R., *Anticipatory Systems: Philosophical, Mathematical & Methodological Foundations*, Pergamon New York, USA (1985).

[54] SAUSSURE, F. de, *Course in General Linguistics*, transl. W. Baskin, The Philosophical Library, originally published 1916 New York, USA (1959).

[55] SEARLE, J. R., "Minds, Brains and Programs", *Behavioral and Brain Sciences* **3** (1980), 417–424.

[56] SEARLE, J. R., *Mind*, Oxford University Press Oxford, UK (2004).

[57] SIEGELMANN, H. T., *Neural Networks and Analog Computation: Beyond the Turing Limit*, Springer Verlag (1999).

[58] SMITH, E. A., *Effects Based Operations: Applying Network Centric Warfare In Peace, Crisis, And War*, CCRP Publication Series (2002).

[59] SOMMERHOFF, G., "The Abstract Characteristics of Living Systems", *Systems Thinking: Selected Readings*, (F. E. EMERY ed.). Penguin Books Harmondsworth, UK (1969).

[60] SUTTON, R. S., and A. G. BARTO, *Reinforcement Learning: An Introduction*, A Bradford Book, The MIT Press Cambridge (1998).

[61] TURING, A. M., "On Computable Numbers, With An Application To The Entscheidungsproblem", *Proceedings of the London Mathematical Society* **42**, 2 (1936-7), 230–265.

[62] VAN GELDER, T., "What might cognition be, if not computation?", *Journal of Philosophy* **92**, 7 (1995), 345–381.

[63] VON ECKARDT, B., *What Is Cognitive Science?*, A Bradford book, The MIT Press Cambridge, USA (1993).

[64] WEINBERG, G. M., *An Introduction to General Systems Thinking* Silver Anniversary ed., Dorset House Publishing New York, USA (2001).

Chapter 24

Wisdom of Crowds in Prisoner's Dilemma Context

Min Sun and Mirsad Hadzikadic
College of Computing and Informatics
University of North Carolina at Charlotte
msun@uncc.edu, mirsad@uncc.edu

James Surowiecki in his book on the wisdom of crowds [1] writes about the decisions made based on the aggregation of information in groups. He argues that under certain circumstances the wisdom of crowds is often better than that of any single member in the group. These circumstances include diversity of opinion, independence, and decentralization. In this paper, we simulated the Prisoner's Dilemma problem as a complex adaptive system, which allowed us to define a "controllable" crowd. Experiments show that in a crowd where the "membership" can be defined dynamically and where members can communicate with each other and learn from each other, the wisdom-of-crowds approach shows advantage over the best performing members in the crowd.

1. Introduction and Background

This paper provides a new way of decision making – using the wisdom of crowds (collective wisdom) [1] to handle continuous decision making problems, especially in a complex and rapidly changing world. By simulating the Prisoner's Dilemma in a Complex Adaptive System (CAS), the key criteria that separate the wise crowd from the irrational one are investigated, and different aggregation strategies are suggested based on different environments.

Decision making has been the subject of research from several perspectives. Generally speaking, decision making is the process of selecting one course of action from several alternative actions. It involves using what you know (or can learn) to get what you want [2]. Since most decisions are personal or individual, which makes it hard to avoid bias, many computer-based Decision Support Systems (DSS) are promoted to help people make decisions in complicated situations for either individual or business purpose. Although knowledge-based decision support systems have been widely used, managers sometimes feel disappointed with their performance because of 1) the difficulties in collecting useful information in a specific field; 2) the cost of setting up and updating knowledge databases; 3) their inherent inadequacies in dealing with complex and rapidly changing environments; and 4) difficulties in determining the proper decision-making model/strategy for problems in social science or economics, which involve numerous human interactions and uncertain personal feelings. With these concerns in mind, a new way of making decisions – using the wisdom of crowds – is introduced, which helps individuals avoid the need for collecting information or setting up and updating knowledge databases, and makes it possible to handle social science or economics problems that involve numerous human interactions and uncertain personal feelings.

In this paper, a simulation using the concept of a Complex Adaptive System is used to demonstrate the wisdom of crowds in the context of the Prisoner's Dilemma problem.

2 Wisdom of Crowds

2.1. Theories of Wisdom of Crowds

A "crowd", in Surowiecki's book, is any group of people who can act collectively to make decisions and solve problems. The wisdom of crowds simply suggests that a collective can solve a problem better than most of the members in the group [1]. This idea appears to be appropriate for explaining behavior of financial markets as expressed by the Nobel-winning economist William Sharpe [3]. It may also be helpful to decision makers in understanding how to solve complex problems. For example, collective voting has already been successfully used by some search engines, including Google [4]. Even though there are many case studies and anecdotes which demonstrate the importance of collective wisdom, there are also many cases supporting the opposite conclusion, some of them cited in "Extraordinary Popular Delusions and the Madness of Crowds," by Charles MacKay [5]. Consequently, not all crowds (groups) are wise; consider, for example, mobs or crazed investors in a stock market bubble. Efforts have been made to understand why and how the wisdom of crowds can take effect. Surowiecki in his book reveals the necessary ingredients, the key criteria which separate wise crowds from irrational ones [1]:

•*Diversity of opinion* -- Each person should have private information even if it's just an eccentric interpretation of the known facts.

•*Independence* -- People's opinions aren't determined by the opinions of those around them.

•*Decentralization* -- People are able to specialize and draw on local knowledge.

•*Aggregation* -- Some mechanism exists for turning private judgments into a collective decision.

2.2. Prisoner's Delimma

Since first raised by Merrill Flood and Melvin Dresher in 1950's [6], a lot of research has been done on the Prisoner's Dilemma (PD) problem, especially after Robert Axelrod introduced the concept of the iterated prisoner's dilemma [7]. PD is a typical type of non-zero-sum game in game theory, based on a well-known expression of PD, the Canonical PD payoff matrix [17], which shows the non-zero net results for the players.

	Player B		
Player A		Cooperate	Defect
	Cooperate	3,3	0,5
	Defect	5,0	1,1

Finding the strategy to gain the highest number of points is the ultimate problem for the Iterated Prisoner's Dilemma. Every year, the IPD tournament [8] is held to evaluate strategies from different competitors. Also, genetic algorithms have been widely used [9, 10] to discover the best strategy. Memory-based strategies and outcome-based strategies such as Tit-For-Tat [11] and Pavlov [11] are regarded as the highly effective ones [12-15].

Extending the "two-player" problem to the "many players" problem brings about the situation where hundreds of players (a crowd) play together. With no central control, players begin to play based on their own strategies. After each round, points are added up for each player. Consequently, a potentially "smart" crowd is formed - the members of the crowd play "cooperate" or "defect" with each other, based on their own strategies. This decentralization of strategies is interpreted as a set of diverse opinions held by the crowd. Then, a simple polling of playing strategies serves as the aggregation method for understanding the vote/wisdom of the crowd.

The crowd in the context of Prisoner's Dilemma satisfies the four key criteria for forming a smart crowd.

a. Diversity and Decentralization

In our Prisoner's Dilemma setting, each agent is given a memory and a strategy. The memory serves to record and accumulate knowledge, thus enabling the agent both to establish a history of games with each player and to accumulate local knowledge. Agent strategies help agents choose whether to cooperate or to defect on the next turn, based on the information in their memory. These strategies also introduce diversity in the game in two ways: generating varying solutions to the problem and drawing conclusions from the local knowledge.

b. Independence

Prisoner's Dilemma in our system (the Iterated Prisoner's Dilemma) allows for the possibility of agents communicating with each other and learning from each other. This, in effect, potentially violates the independence criterion of Surowiecki. This aspect is fundamentally different from the cases Surowiecki described in his book.

However, we also introduced a switch for independence in the system, which enables us to conduct experiments with both independence and learning-evolved environments.

c. Aggregation

A statistical or social science aggregation means combining outputs/solutions from different lower-level entities into summary indicators. In the Prisoner's Dilemma Problem, aggregation characterizes the group-level perormance by combining the individual member's contributions (or solutions), regardless of whether these contributions are duplicated, contradicted, or incomplete. The most commonly used methods for this type of aggregation are sampling, polling, and voting.

2.3. Prisoner's Dilemma in CAS

Complex Adaptive Systems represent a dynamic network of agents (which may represent cells, species, individuals, firms, nations, etc.) working in parallel, constantly acting and reacting to what other agents are doing [16,17].

We extend the two-player Prisoner's Dilemma game into a situation involving hundreds of players (crowd) playing against each other in a pairwise fashion. Agents play against each other repeatedly without a central control. It is natural to describe the Prisoner's Dilemma as a complex adaptive system in order to reveal spontaneous reactions among individual players, as well as the wisdom hidden inside the group as a whole. This allows for exploration of various aggregation strategies.

3. Experiment Results

3.1. Implementation

In order to design a CAS for Prisoner's Dilemma, first we need to create: 1) individual "player-agents" who can "cooperate" or "defect" when playing the game based on their own strategy; and 2) special "aggregator-agents" who use the wisdom of crowds by acting as aggregators of various groups within the crowd. Since agents play against each other repeatedly without the central control, we assign each agent a memory that is used to store information (knowledge) about previous "matches". The player-agents initially "receive" a strategy which they use to decide their actions based on the information they have, while the aggregator-agents are given the ability to make their decisions by consulting with their "advisory group," formed from the set of player-agents selected by each aggregator-agent.

In the system each player-agent is described using a chromosome-like structure [18]:

Agent Number	Basic Strategy	Limitation	Reaction1	Reaction2

Where:
- *Agent Number* is the number used to identify each player.
- *Basic Strategy* is the number indicating the strategy an agent chooses to guide its behavior.

- *Limitation* is a number that modifies the Basic Strategy, as described below. Combined together, these two numbers define the judgment of the situation the agent is facing.
- *Reaction1* defines the behavior of the agent if the situation described by Basic Strategy + Limitation applies in the current case/match.
- *Reaction2* defines the behavior of the agent if the situation described by Basic Strategy + Limitation does not apply in the current case/match.

Aggregator-agents represent special participants (competitors) in the game. On each turn, Aggregator-agents choose to cooperate or defect according to the opinions from their chosen People-agent group. An Aggregator-agent has no strategy which can give it guidance for cooperating or defecting; its only strategy is to decide which People-agent group it wants to listen to and the manner of aggregating the group's advice.

Each Aggregator-agent is described using a chromosome-like structure:

Agent Number	Selection Strategy	Select Number	Aggregation Strategy

Where:
- *Agent Number* is the number used to identify each Aggregator-agent.
- *Selection Strategy* is the number indicating the strategy used to select a People-agent group.
- *Select_Number* is the number indicating how many People-agent are chosen to form the group; it can be any number from 1 to the maximum player-agent number.
- *Aggregation Strategy* is the number indicating the strategy used for aggregating from the selected group.

In the implementation, a set of basic and/or selection strategies is assigned randomly to each agent. If two agents meet, a match is initiated. Agents play with each other according to the strategy they selected and the information they have about each other. After each play, the points are added to the players and the agents move on to the next match. After running the simulation for T times all player-agents are evaluated using the fitness function based on the overall points collected. During the simulation, poor performers are replaced by the best performer if that strategy is selected; otherwise, player-agents with higher scores are replaced. In addition, a lesser performer can opt to copy the whole or part of the better player's strategy.

3.2. Experiment Results

Experiment 1: Player-agents' and Aggregator-agents' performance in both no-learning crowds and evolutionary crowds

Adding the learning ability to the player-agents enables them to learn individually to improve their décisions. Although this violates part of Surowiecki's criteria –independence – it is common and necessary in real life. Experiments show that by keeping enough diversity of opinion, the aggregate wisdom of the crowd can still perform better than most individual members, even better than the best individual. In this experiment, we focus on the agents' performance both with and without the ability of crowds learn.

Since the formation of crowds is important to the performance of agents in the Prisoner's Dilemma Problem, we run the experiments 10 times with different random seeds. During each round, 250 player-agents are placed in the game. After player-agents have had a chance to play against each other and learn from each other for certain learning period, aggregator-agents with different strategies are introduced into the game. Aggre1, 5, 9.., 250 represent the aggregator-agents with differing aggregation strategies. For example, an aggregator-agent whose strategy is to consult player-agents with the highest scores may choose to follow the advice of the group of player-agents having the current highest score, and we call it aggre1, likely a wise strategy for the aggregator-agents. Similarly, best_people, median_people, average_people represent the player-agents. For example, best_people represents one of the player-agent having sthe current highest score.

In Figure 1, the charts show the performance of player-agents and aggregator-agents, after certain duration of learning, using 10 different seeds (formations of crowds).

By introducing the ability to learn, the performance of player-agents and Aggregator-agents show increased volatility in terms of scoring for different seeds (crowds). When no learning is introduced, the performance of player-agents and aggregator-agent is relatively stable, no matter which seed (formation of crowds) is used. Although the line for best-people is always on the top in Figures 1 and 2, it is interesting to notice that the lines for Aggre19 and Aggre29 are close to the one for best_people, which suggests that the best way to make decisions using the wisdom of the crowd, in this situation, is to listen to the 10% best individuals in the crowd, so that the performance will be similar to the best individual in the crowds but only slightly lower. The Best individual's performance might change with each tick, but the performance of aggregators remains solid all the time.

When we introduce a learning period more volatility occurs, and best-people agent is no longer the all-time winner. In Figure 1, which shows the situation after Learning was turned on for 150,000 ticks, the aggregator-player performs better than the best_people six times out of ten. This suggests that more than half the time, making a decision using the wisdom of the crowd is even better than the best individual in the crowds.

Experiment2: Player-agents' and Aggregator-agents' performance varying in the size of crowds

The size of crowds is another factor influencing agents' performance. In this experiment, we focus on the player-agents' and aggregator-agents' performance, while varying the size of the crowds. Two sets of experiments were run using different random seeds: 250 player-agents and 500 player-agents.

In Figure 2, avg_people_250 and best_people_250 represent the average player-agent and the player-agent with the current high score in a crowd of 250 player-agents; aggre250 represents the aggregator-agent who chooses the strategy to listen to all 250 player-agents in the crowd; likewise for avg_people_500, best_people_500 and aggre500.

The charts in Figure 2 show that despite using various random seeds, the increased size of the crowd (which increases the diversity of opinion) results in a better performance for both player-agents and aggregator-agents. In addition, the aggregator-agent using the wisdom of the crowd approach performs better most of the time than the best player-agents in those crowds.

Wisdom of Crowds(C) VS Best People-agent (B)VS Average People-agent (A)

# out of 10	C > B > A	B > C > A	B > A > C
250 agents	6	2	2
500 agents	6	3	1

4. Lessons Learned and Future Work

In the previous sections, we extended the concept of Wisdom of Crowds to a continuous decision making problem – The Prisoner's Dilemma. A simulation using the concept of Complex Adaptive Systems was built to demonstrate the concept of wisdom of crowds, while at the same time evaluating Surowiecki's four criteria for forming a smart crowd. However, it is hard to imagine a continuous decision-making example where members of the crowd are truly independent from each other in the real world. Therefore, by partially violating the independence criteria, we added the learning ability to the crowd. Our experiments show that this addition makes both individual players and the aggregate-players smarter, while still guaranteeing the diversity of opinion. Furthermore, these experiments show that in a crowd where the "membership" can be defined dynamically, and where members can communicate with each other and learn from each other, the wisdom-of-crowds approach is superior to the best performing members in the crowd.

Our future work will help us to: 1) Characterize the structure of crowds more precisely way using elements such as size and density ; 2) Identify the behavior of the crowds with different agent settings: heuristic, behavior pattern, social influence, and learning speed; and 3) Quantify and qualify the characteristics of aggregators.

Bibliography

[1] James Surowiecki, 2004, *"The Wisdom of Crowds: Why the Many Are Smarter Than the Few and How Collective Wisdom Shapes Business, Economies, Societies and Nations"* ISBN 0-385-50386-5

[2] Walker, Doris Katey, 1987, *"Improving Decision-Making Skills"* Manhattan, KS: Kansas State University Cooperative Extension Service, MF-873

[3] Ayse Ferliel, 2004, *Inteview with William Sharpe*, Investment Adviser

[4] David Austin, 2006, *"How Google Finds Your Needle in the Web's Haystack"*, American Mathematical Society Feature Column

[5] Charles MacKay, 1841, *"Extraordinary Popular Delusion and the Madness of Crowds"*, ISBN 1853263494, 9781853263491

[6] Flood, M.M., 1958, *"Some Experimental Games, Research Memorandum"* RM-789,RAND Corporation

[7] Robert Axelrod, 1984, *"The Evolution of Cooperation"* New York: Basic Books ISBN 0-465-02121-2

[8] http://www.prisoners- dilemma.com/results/cec04/ipd_cec04_full_run.html

[9] Axelrod, R, 1987, *"Evolving New Strategies"*, Genetic Algorithm and Simulated Annealing, ISBN-10: 0934613443

[10] Jennifer Golbeck, 2002, *"Evolving Strategies for the Prisoner's Dilemma"*, Advances in Intelligent Systems, Fuzzy System, and Evolutionary Computation

[11] http://www.iterated-prisoners-dilemma.net/prisoners-dilemma-strategies.shtml

[12] Fogel D., 1993, *"Evolving Behaviors in the Iterated Prisoner's Dilemma"*, Evolutionary Computation 1(1)

[13] Darwen, P. and X, Yao, 1994, *"On Evolving Robust Strategies for Iterated Prisoner's Dilemma"*, Lecture Notes in Artificial Intelligence, AI '93 and AI '94 Workshops on Evolutionary Computations, Melbourne, Australia

[14] Kraines D and Kraines V, 1993, *"Learning to Cooperate with Pavlov – an adaptive strategy for the Iterated Prisoner's Dilemma with Noise"*, Theory and Decision

[15] Kraines D and Kraines V, 1995, *"Evolution of Learning among Pavlov Strategies in a Competitive Environment with Noise"*, Journal of Conflict Resolution

[16] M. Mitchell Waldrop, 1992, *Complexity: The Emerging Science at the Edge of Order and Chaos*

[17] http://en.wikipedia.org/wiki/Complex_adaptive_system

[18] Mirsad Hadzikadic and Min Sun, 2007, *"A CAS for Finding the Best Strategy for Prisoner's Dilemma"*, ICCS 2007

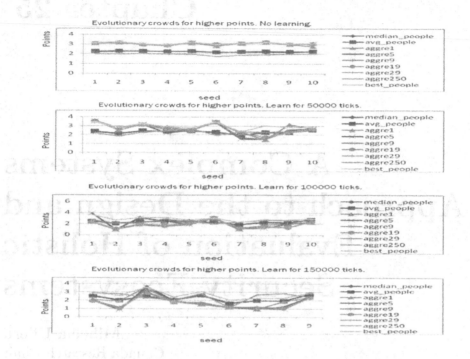

Figure 1: Performance of agents in Different Seeds, Varying in Duration of Learning

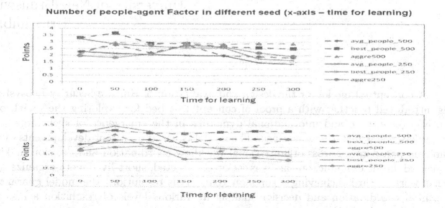

Figure 2: *Performance of agents, varying in Size of Crowds*

273

Chapter 25

A Complex Systems Approach to the Design and Evaluation of Holistic Security Ecosystems

Mihaela Ulieru
Canada Research Chair
Faculty of Computer Science
University of New Brunswick
ulieru@unb.ca

A conceptual model for the design and evaluation of Holistic Security Ecosystems is introduced together with a proof of concept test bed for exploring the social, cognitive, geographic and informational dynamics in the emergence of such large scale interdependent meta-organizations. The model is based on recent developments of the holonic paradigm of cross-organizational workflow coordination and decision making. Focused on the dynamic, on-the-fly creation of targeted, short-lived meta-organizations that work towards achieving a common goal (crisis resolution) the model guarantees optimal coordination and decision making at various levels of resolution across the holarchic levels of the organization.
Keywords. Holonic Enterprise, Command and Control (C2), complex interdependent networks, Complex Adaptive Systems (CAS), network-enabled operations, Emergence of Robust Structure.

1 Introduction

The new security challenges of the 21st century are qualitatively different than in the past. Due to the complexity of such operations military forces find themselves collaborating with numerous other 'partner' organisations to achieve a common goal. This has imposed new demands on capacities and capabilities; and consequently requires new models to understand key issues and evaluate options. Successful modeling must consider human and organisational factors, which are currently not adequately addressed; certainly not at a meta-organisational level. The problems the military forces and their 'partners' encounter in operations are often messy, intractable, and dynamic; spilling across the problem-solving and management boundaries of single organisations or established cross-sector forms. The environment may often border on the chaotic and uncontrollable, but it may be possible to influence it in a predetermined direction if approached properly. This will require effective collaborative problem solving on the part of partnering organisations for which their "common goal" is more accurately a commonality of elements which bind them together in collective action. It is generally the intent of these partnering organisations to retain their autonomy while "joining forces" to achieve shared goals. The resulting tensions between autonomy and partnering lead to ambiguity and complexity in the meta-organisational (i.e. the collective set of entity organisations and interrelationships) structure or form. These tensions must be reconciled in order to achieve both individual and shared objectives. Participants are pushed into activities that are beyond traditional areas of competence and they are stressed when encouraged simultaneously to build inter-organisational linkages and to protect organisational autonomy. In these instances, both cooperative and competitive behaviour will likely be observed. The persistence of "coordination" as a problem in operations indicates a deeper issue than merely the need to "coordinate" tasks, which relates to the nature of the relationships amongst entities within a meta-organisation and whether or not the set of relationships and consequent meta-organisational form promotes or hinders collective decision-making. In recognition of this problem, theories on "robust networking" have been advanced but require not only shared information but shared understanding and intent as well. It is rarely argued any longer that technology drives social change; instead a more holistic approach is advocated in which "information technology" is comprised not only of physical artefacts but also the social relations around those artefacts.

In this paper we propose a complex systems approach to the emergence of holonic organizational structure – as meta-organizational structure (system of systems) integrating various players (multi-level organizations, individuals, devices and the ICT systems and communication networks linking them) while balancing autonomy and cooperation in the drive towards a common goal (crisis resolution) in emergency response-related military operations.

2 On Emerging Robust Structure Through Architecture and Protocols

For a decentralized organization to function as an organization-and not just as a collection of disconnected elements – the components must interact within a shared environment, typically internal to the organization. The components

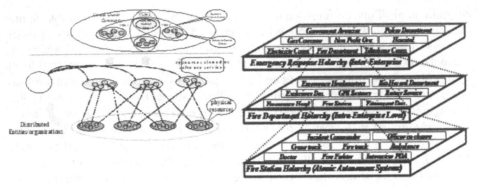

(a) Reconfigurable Holarchy with Dynamic Mediator

(b) Emergency Response Holarchy

Figure 1: Emerging Robust Holonic Structure

can no longer simply report up their respective chains of command and expect insightful decisions to issue magically from the top. The components must interact among themselves and find their own ways of collaborating through the environment.

Command and Control (C2) is the military term for the structures and processes though which an entity (i.e., an organization, a system, an organism, etc.) operates. Every entity (military, business, social, political, biological, hardware, software) has a C2 structure. Much of an entity's C2 structure is often recorded in its constitution, by-laws, policies and practices manuals, or design documentation - if it has any of these. Virtually no entity has a complete statement of its C2 structure. A fundamental question to be faced by any discussion of C2 is: what are the requirements for which a C2 design is the answer. That is, what meta-behavioral properties or qualities do we want an organization/system to have? Among the list of possible requirements are the ability to choose actions which will further the system's interests, the ability to act effectively to perform a specific function (sometimes known as execution), the ability to respond to the unknown, the ability to act at the appropriate time scale depending on the situation, the ability to recover from injuries, etc.

From a complex systems perspective, C2 can be built into the architectural requirements determining the components and their interactions through protocols encapsulating the policies and governance rules, which thus will shape the structure of such an organization. Governance refers to the creating of conditions for ordered rule and collective action [9], and focuses on centrally controlling major societal functions to reduce chaos and preserve overall system optimal functionality with respect to all its participants. The C2 mechanism lays the foundation for emerging robust structure [10] in the timely deployment of dynamic, short-living organizational structures needed in emergency response military operations, Fig. 1 a. b. Robustness stems from system's ability to reconfigure its structure to accommodate various disturbances while maintaining

276

its functionality in a range of acceptable behaviors. This is achieved via a 'plug-and-play' flexible architecture in which components can be easily interchanged to take over the functionality of broken ones [13].

[1] makes a strong argument regarding the role of architecture and protocols in the evolution of complex systems - in particular in the capacity to develop resilience through robust structure. Protocols define how diverse modules interact and architecture defines how sets of protocols are organized. The concepts of architecture and protocol is completely compatible with the challenge of developing new ways to organize human effort beyond the classic industrial control hierarchy. While we have tend to explore inter-organizational architectures for collaboration across a wide range of efforts, there have been far few efforts to explore the architectural design space within an organization. Our standard architectural framework has been the control hierarchy and protocol is hierarchic authorizations.

Doyle's deep analysis of biological and technological robustness concludes that selection acting at the protocol level could evolve and preserve shared architecture, thus enabling interchangeable 'plug-and-play' of components, which in turn facilitates structural reconfiguration. This is in tune with our previous result [10] while deepening and fleshing out what makes auto-catalytic sets, the fundamental units of self-reproductible complex systems architecture [7]. The basic auto-catalytic set (holon) within the underlying architecture/protocol lays the foundation for emerging (robust) structure and preserving it during the dynamics of purposeful organizational deployment in the chaos of crisis [11].

Ecologies and economies framed by suitable architecture components are defined by the workings of the protocol allowing integration of components parts. Mastering protocol and architectural design at the 'primal' autocatalytic set / holon is thus crucial in ensuring resilient deployment, given that it is the protocol that creates the dynamics of organizational boundary. Resilience of a social ecosystem is defined as the capacity of the system to absorb disturbances while maintaining its function, structure, identity and feedbacks [15]. Resilience depends the capacity of the organization to re-organize over spatial and functional scale [4] via adaptive governance [2].

The participants in a military operation may be described as species within a social ecosystem [3] specialized to achieve both their own goals and those of the greater organization [4]. Such organizations are characterized by:

- the participants' ability to negotiate between autonomy and cooperation in a drive (attractor) towards a common goal,

- a coordinated workflow process that triggers the formation of high-level organizational structure (patterns of collaborative clusters) through low-level interactions between participants, and

- a capacity to organize over spatial and functional scale [15] to maintain resilience against attack.

3   Holistic Security Ecosystems (HSE)

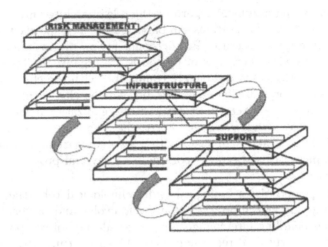

Figure 2: Holistic Security Ecosystem

We build on the holonic enterprise and emergency response holarchy concepts, Fig. 1 to define a *holistic security ecosystem (HSE)* as an emergent short-living meta-organization dynamically created in response to an emergency event by bringing together several otherwise stand-alone dispersed organizations [14]. The HSE is a meta-organization of interdependent specialized Risk, Support ad Infrastructure Holarchies, Fig. 2 working in synergy through a shared environment – most fundamentally a communication network - which adds one more dimension (C) to the Command and Control – making the operational coordination across an HSE a C3. C3 is facilitated by a shared environment, including common resources as well as implicit and explicit rules of behavior. Management of the interactions between these organizations has to undertake multifaceted challenges (cultural, professional, coopetition, trust in a new temporary authority, etc) which require careful crafting of the basic architecture and protocol elements to enable resilient flexible functionality in an unpredictable dynamic environment. Such an organization is subject to either gradual or abrupt change. Gradual change is characterized by a steady progression in organizational change, whereas abrupt change is characterized by unpredictable actions and consequences [5]. In the case of an attack, periods of abrupt change increase in frequency, duration and magnitude.

To increase the flexibility of military units approaches such as net-centricity have been proposed, which imply a significant decentralization of authority – individual components of an organization are given as much autonomy as possible. Yet virtually all organizations remain hierarchical to some degree, thus the holonic heterarchical structure suits well the purpose of balancing autonomy of low-level holons with the authority of a chief executive / unity of command encapsulated in the HSE via a dynamic mediator [13] Fig. 3 enabling authority to be dynamically allocated at various levels in the chain of command as well as within one level (in case e.g. the chief executive needs to be replaced). The executive (mediator) is given the authority and responsibility to use some assigned resource(s) – typically more than s/he can control on his or her own-to achieve some objective.

One may look at an organization's operational structure as a reflection of its strategy for allocating resources. This opens the perspective of using market models to reconfigure the organizational structure via the harmonious flow of resource allocation tuned to respond optimally to the crisis at hand. Markets (and most innovative environments) allocate resources in a bottom-up fashion. It is primarily the autonomous agents that decide how resources will be distributed. They make that decision when they make their individual decisions about what to buy. Similarly, "power to the edge" implies that the power to allocate resources is vested primarily with the lowest level elements-those at the edge, away from the power centers. This approach opens the possibility of tuning the bottom-up emergence of robust structure via market models [8].

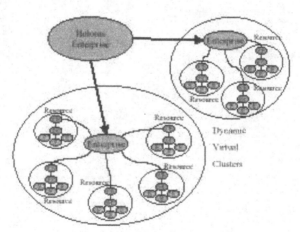

Figure 1 - Dynamic Virtual Clustering Pattern
in the Holonic Enterprise

Figure 3: Delegation of Command via Dynamic Mediator

When tuning the resilience of HSEs via resource allocation it is important to realize that entities capable to acquire their own resources from sources outside of themselves (and from outside of any larger organization of which they are a component) can be far more autonomous than entities that acquire their resources from higher levels within a hierarchical resource allocation framework. Thus, a niche for independence/autonomy in a holarchy can be created by outside suppliers which will thus create a buffer accommodating eventual resource scarcity strains that may lead to cascading failures otherwise. Such external sources un-

279

dertaking eventual unexpected loads in case of unexpected disturbances enforce
organization's resilience.

4   Modeling and Simulation Testbed

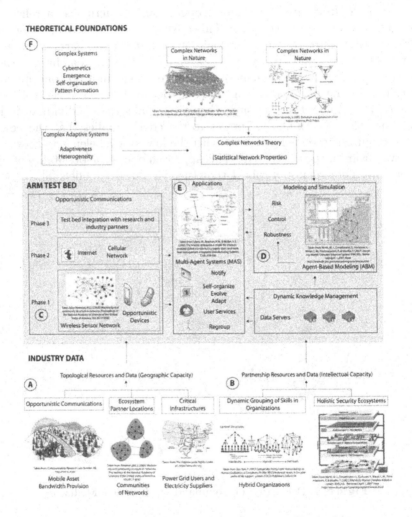

Figure 4: Adaptive Risk Management Testbed

Among the major challenges facing the deployment of such dynamic interdependent meta-organizations, we mention: How are decisions made about both allocating existing shared resources and investing in new shared resources? How to craft rules that govern both behavior and the use of shared resources? Once made, how are these rules enforced? How are they changed as circumstances change? To address these challenges we are working on the development of a conceptual model for the emergence (dynamic creation) of HSE via collaborative resource exchange among participants. This boils down to the modeling and analysis of interdependent network-enabled hybrid complex systems consisting of organizations, departments, individuals, information and physical entities and the dynamics of their cascading effects under various conditions and strains. Simulations on the adaptive risk management (ARM) testbed available in our lab, Fig. 4 [12] enable an understanding of the dynamics of criticality occurrence within the Holistic Security Ecosystem for a wide range of operating scenarios.

The conceptual model (B in fig. 4) encompasses two capacities:

- The **geographical capacity** of the organization addresses which resources ("partners") are located where at any given time. On our testbed the geographical placement of organizational partners is modeled through the Wireless Sensor Network (C in Fig. 4), where every sensor represents the location of a collaborative partner.

- The **intellectual capacity** of the organization consists of the specialized skills available through different partners in the organization. An indication of responsiveness, focus area of employees etc. would be typical examples of intellectual capacities.

The geographical intellectual capacities represent the organization and its partners as a network, whose entities are processing by the modeling and simulation module (D in Fig. 4).

The HSE Testbed (Fig. 5) is being used to run various configurations of HSE under various conditions and strains with various factors impacting the workflow coordination and decision-making throughout the meta-organization to enable understanding of the dynamics of criticality occurrence within the HSE under various operating conditions / scenarios of mission critical activities. Through simulations, existing social networks are "mapped" into the holonic model to investigate the strengths and resilience of various HSE configurations, thus determining their suitability to address various crisis models. This enables mapping of various HSE configurations to the crisis types for which the particular meta-organizational structure works best. Validation of resulted HSE configuration – crisis type mappings on 'in-vivo' simulation exercises for various instantiations of scenarios (taking e.g. pandemic mitigation or Vancouver Olympics scenarios an various crisis possibilities within the particular scenarios) will provides essential feedback for the model improvement.

5 Conclusions and Future Work

We will extend the holonic model by integration of various aspects impacting the flow of decision-making and functionality of the meta-organization (professional decision-making, cultural impact, trust in such short-life mission-oriented organizations, etc.). Analysis and identification of the impact and interdependencies between various key factors in the extended model transcending cul-

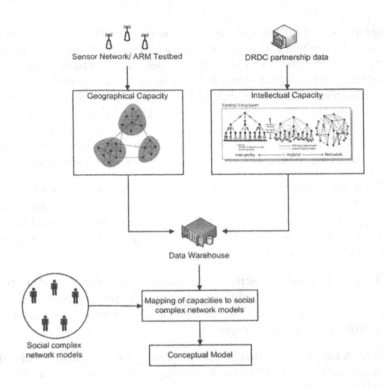

Figure 5: Holistic Security Ecosystem Simulation Testbed

tural, professional, psychological, sociological etc. dimensions will be used to tune the resilience of the HSE by identifying eventual cascading effects with emphasis on the weakest points/links, to determine counteracting (strengthening) measures. This will result in a methodology of design for resilience of HSE laying the foundation of a template for harmonious inter-organisational operations coordination in highly dynamic, short-living mission-critical crisis relieving meta-organizations encompassing methods to optimise interactions and communication linkages among participants.

Integrating the simulations results into a 'strategic thinking process' will enable a change of culture in the design and deployment of integrated HSE (with a-priori identified risks and potential cascading criticalities strengthened and an anticipatory ability of the impact of various dynamics of interdependent factors) which would lead to a seamless reorganization of the HSE in patterns of resilience under various strains and internal disturbances – that will enable it to keeping its operational flow unobstructed through the chaos of various crises. If taken to the next level – this could lead to an overall benchmarking of strategic thinking for self-transformation to help organizations adapt to the high dynamics of our world by considering interdependent factors while better focusing on relevant strength in overcoming limitations, [6].

We must ensure that today's solutions are not tomorrow's problems - and

282

key to this is our capacity of agile response directed by wise strategy. We hope that the proposed conceptual model and testbed will facilitate 'wise strategy' deployment by crafting emergent robust structure in dynamic meta-organizations.

Bibliography

[1] CARLSON, J.M., and J. DOYLE, "Complexity and robustness", *Proceedings of the National Academy of Science USA 99, Suppl 1*, (2002), 2538–2545.

[2] DIETZ, T., E. OSTROM, and P.C. STEM, "The struggle to govern the commons", *Science* **302** (2003), 1907–1912.

[3] FOLKE, C., S. CARPENTER, B. WALKER, M. SCHEFFER, T. ELMQVIST, and L. Gunderson et AL, "Regime shifts, resilience and biodiversity in ecosystem management", *Annual Review in Ecology Evolution and Systematics* **35** (2004), 557–581.

[4] FOLKE, C., T. HAHN, P. OLSSON, and J. NORBERG, "Adaptive governance of social-ecological systems", *Annual Review of Environment and Resources* **30** (2005), 441–473.

[5] GUNDERSON, L.H., and C.S. HOLLING, *Panarchy: Understanding Transformations in Human and Natural Systems*, Island Press, Washington D.C. (2002).

[6] HOLLAND, J.H., *Hidden Order: How Adaptation Builds Complexity*, Addison-Wesley, Reading, Mass. (1995).

[7] KAUFFMAN, S., *Investigations*, University Press, Oxford (2000).

[8] NORTH, M.J., G. CONZELMANN, V. KORITAROV, C.M. MACAL, P. THIMMAPURAN, and T. VESELKA, "E-laboratories: Agent-based modeling of electricity markets", *American Power Conference, Chicago, IL*, (2002).

[9] STOKER, G., "Governance as theory: Five proportions", *International Social Science Journal* **50(1)** (1998), 17+.

[10] ULIERU, M., "Emerging computing for the industry: Agents, self-organization and holonic systems", *Proceedings of the IEEE-IECON International Workshop on Industrial Informatics*, (2004), 215–232.

[11] ULIERU, M., "Adaptive information infrastructure for the e-society", *Engineering Self-Organizing Systems: Methods and Applications*, (SERUGENDO AND KARAGEORGIOS eds.). Springer, Berlin (2005), pp. 32–51.

[12] ULIERU, M., and S. GROBBELAAR, "Engineering industrial ecosystems in a networked world, keynote paper", *5th International Conference on Industrial Informatics*, (2007).

[13] ULIERU, M., and D. NORRIE, "Fault recovery in distributed manufacturing systems by emergent holonic re-configuration: A fuzzy multi-agent modeling appraoch", *Information Science* **7669, September** (2000), 101–125.

[14] ULIERU, M., and P. WORTHINGTON, "Adaptive risk management systems-for critical infrastructure protection", *Integrated Computer-Aided Engineering* **13:1** (2006), 63–80.

[15] WALKER, B., C.S. HOLLIN, S.R. CARPENTER, and A. KINZIG, "Resilience, adaptability and transformability in social-ecological systems", *Ecology and Society* **9(2)** (2004), –.

Chapter 26

Exploring Watts' Cascade Boundary

Daniel E Whitney
Massachusetts Institute of Technology
dwhitney@mit.edu

Watts' "Simple Model of Global Cascades on Random Networks" used percolation theory to derive conditions under which a small trigger causes a finite fraction of an infinite number of nodes to flip from "off" to "on" based on the states of their neighbors according to a simple threshold rule. The lower boundary of this region (at approximately $z = 1$) was determined by the disintegration of the network while the upper boundary (at values of z that depend on the threshold) was determined by the resistance of nodes to being flipped due to their many connections to stable nodes. Experiments on finite networks revealed a similar upper boundary, displaced upward from the theoretical boundary toward larger values of z. In this paper we study cascades on finite networks in this upper boundary region via simulations. We distinguish two kinds of cascades: *total network cascades* (or TNCs) that essentially consume the entire finite network, and cascades corresponding to those predicted by percolation theory that consume only the initially struck vulnerable clusters and possibly a few others. We show that the experimental upper boundary found by Watts corresponds to TNCs. TNCs can start when, for example, as few as 2 vulnerable nodes in a network of 10000 nodes are flipped initially by a single seed and no cluster of vulnerable nodes larger than 21 exists. The mechanism by which these cascades start and grow near the upper boundary is not described by percolation theory because the nodes involved form a densely connected subnetwork that is not tree-like. Instead a different mechanism is involved in which a particular motif comprising patterns of linkages between vulnerable and stable nodes must be present in sufficient quantity to allow the cascade to hop from the initially struck vulnerable clusters to others. While the cluster-hopping mechanism is necessary to start the cascade, the later emergence of a TNC requires the presence of relatively large clusters of vulnerable nodes. The shrinkage of the size of these largest clusters relative to network size is a major factor in the disappearance of TNCs as finite networks with the same z get larger. TNCs are also enhanced if the network is artificially altered by degree-preserving rewiring to have positive degree correlation or increased clustering coefficient. Each enhances the likelihood of TNCs by a different mechanism. Great variability is observed in nominally identical random networks with the same z with respect to properties such as the amount and distribution of the cluster-hopping motif and the size of the largest vulnerable cluster. This means that metrics based on first moments of these or other characteristics are unlikely to reveal which networks are susceptible to TNCs.

I. Introduction

Watts [Watts] used percolation theory to model the ways rumors or other influences might propagate through populations based on a simple threshold rule: Each node is assigned a threshold ϕ and an initial state 0 or "off." If a fraction ϕ or more of a node's neighbors acquire state 1 or "on," then that node will flip from off to on. The threshold corresponds to a critical nodal degree $K*$ such that $K* = \lfloor 1/\phi \rfloor$. Nodes with $k \le K*$ are termed "vulnerable" and will flip if one of their neighbors flips. Nodes with $K* < k \le 2K*$ are here called "first-order-stable" and will flip if two of their neighbors flip. Nodes with $2K* < k \le 3K*$ are called second-order-stable and will flip of three if their neighbors flip, etc. Watts found exact solutions for infinite Poisson random graphs that define a region in $z - \phi$ space (where z is the network's average nodal degree) inside of which a finite fraction of an infinite network would flip from off to on if even one node (called the seed) were arbitrarily chosen and flipped from off to on. (Figure 1) He performed simulations on finite networks comprising 10000 nodes and found that these networks exhibited a similar region. The lower boundary of this region (at approximately $z = 1$) was determined by the disintegration of the network while the upper boundary (at values of z that depend on the threshold) was determined by the resistance of nodes to being flipped due to their many connections to stable nodes.

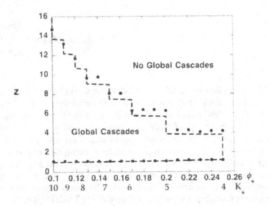

Figure 1. The Cascade Window [Watts]. Black dots correspond to rare global cascades observed in random networks containing 10000 nodes.

In this paper we examine the region in finite networks where infinite networks have their upper boundary and seek understand why global cascades (called total network cascades or TNCs in this paper) occur in what appears to be an infertile region. This is interesting for several reasons. First, percolation theory speaks only to the flipping of vulnerable nodes, so the flipping of stable nodes is accomplished, as Watts observes, by numbers of flipped vulnerable nodes surrounding stable nodes and flipping them. But this mechanism operates only when a cascade is mature, not when it is just starting. Second, the absolute size of the cascade is, in percolation theory, upper-bounded by the number of vulnerable nodes, but the entire network can be made to flip even when vulnerable nodes comprise a small fraction of the entire network and clusters of them are small. Third, much of the interesting commentary Watts gives pertains to the finite networks on which simulations were performed, not infinite networks. These facts and statements prompted the present study.

II. Characteristics of Networks Near the Upper Boundary

The networks being studied comprise two kinds of nodes, namely vulnerable and stable. Each kind appears in clusters of various sizes, depending on z. When z has a value in the middle of the region, most of the nodes are vulnerable and belong to one large cluster, while the stable nodes are dispersed in many smaller clusters. Near the upper boundary most of the nodes are stable and belong to one large cluster while it is the vulnerable nodes that are dispersed in many smaller clusters. A typical situation contrasting clustering in the middle of the cascade window and above the theoretical upper boundary is shown in Figure 2.

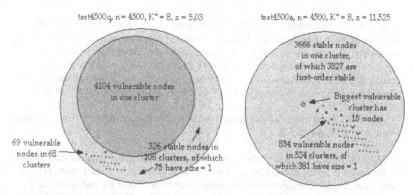

Figure 2. Clustering in Two Finite Networks with $n = 4500$ and $K^* = 8$. Left: In the middle of the cascade window where network test4500q has $z = 5.03$. Right: Above the theoretical upper boundary where network test4500a has $z = 11.525$. TNCs occur, though rarely, in this network.

Even above the theoretical boundary, there is no shortage of vulnerable nodes: when $K^* = 5$ and $z = 6.4$, typically 37% of the nodes are vulnerable. When $K^* = 8$ and $z = 11.6$, typically 18% are vulnerable. In both cases, 90% of the stable nodes are first-order stable, meaning that they will flip if two neighbors flip. Thus cascades do not become rare because stable nodes are so highly linked that there is little chance of overcoming their threshold.

One might wonder why it is then so hard to start cascades in this region. According to percolation theory, what is happening is that the vulnerable nodes are becoming isolated from each other so that clusters of them are becoming smaller as z increases for a given K^*. Finite networks display analogous behavior. While the sizes of the average and largest vulnerable cluster increase as the size of the network increases for given z and K^*, the largest vulnerable cluster size scales approximately as $n^{0.27}$ when $z = 11.5$ and $K^* = 8$ so that in an infinite network the size of the largest vulnerable cluster would be zero.[1,2] Typically, 45% or more of the vulnerable nodes are in clusters of size = 1, meaning that they are isolated from other vulnerable nodes.

---

[1] The second moment of vulnerable cluster size may be taken as a proxy for the size of the largest cluster. This can be calculated in the random network case as $H_o''(1)$. This quantity diverges at the same value of z as does the first moment, so the behavior of this quantity is no different than that of mean cluster size.

[2] The size of the largest vulnerable cluster is not an outlier. The largest one is followed by many that are almost as large. As discussed later in the paper, large vulnerable clusters are strong enablers of TNCs.

Since large clusters of vulnerable nodes[3] are rare or nonexistent in the region just above the theoretical upper boundary, and since percolation theory speaks only to the growth of cascades within these small vulnerable clusters, how do TNCs get started? The experiments described below were aimed at answering this question.

III. Experiments

In this study, simulations were performed on networks of various sizes from 2000 nodes to 36000 nodes, following Watts' procedure [Watts b][4]: a value of $K*$ and a value of z were chosen such that z was larger than the theoretical value for that $K*$ but at or near the boundary value found by Watts; a random network was generated having approximately that value of z; a single seed node was chosen at random and flipped from off to on. Subsequently, one of the following Scenarios occurred:

1. The seed did not link to any vulnerable node and nothing happened. This occurred in most of the experiments.

2. The seed linked to one or more vulnerable nodes; these nodes flipped, then the rest of the nodes in their clusters flipped (that is, their clusters percolated), then (or concurrently) occasionally vulnerable nodes in another cluster flipped, and then the cascade stopped after flipping at most a few dozen nodes. This event occurred often, perhaps once in 10 tries.

3. As in 2, but the entire network flipped. This event was rare, occurring once in several hundred tries, and more rarely for larger networks having the same z.

Scenario 2 is completely predictable in infinite networks by percolation theory as applied by Watts. Its occurrence in finite networks obeys similar conditions, namely that some non-zero-size vulnerable clusters exist and that the seed link to one or more of them. The sizes of such cascades are distributed exponentially, as determined experimentally by Watts. The sizes of vulnerable clusters in finite networks are also distributed exponentially, determined experimentally here. This fact permits us to surmise that the exponentially distributed cascades whose sizes are plotted by Watts in his Figure 3 are Scenario 2 events. The upper limit of z above which Scenario 2 can no longer occur is that where the largest vulnerable clusters consist of a single node each. This value of z is far higher than the theoretical value at the upper cascade boundary in infinite networks or the location of the black dots in Watts' Figure 1. In fact, as determined experimentally for this paper, the upper limit for Scenario 3 in finite networks corresponds to the black dots. The one data point in Watts' Figure 3 corresponding to a cascade consuming 100% of the network then presumably is a Scenario 3 event. Note that the vulnerable clusters in finite networks near the theoretical upper boundary, while large enough to account for Scenario 2 events, are too small to comprise TNCs, and the totality of vulnerable nodes is no more than 20% of the network when $z = 11.5$ and $K* = 8$.

---

[3] Note that we repeatedly refer to "large vulnerable clusters" or "relatively large vulnerable clusters" in this paper. In no case are such clusters larger than about 1% of the size of the network, and usually they are smaller. But the size of the largest vulnerable cluster is important in determining whether a cascade, once launched, can become a TNC.

[4] Watts generated a new network for each of his experiments. In this paper we ran upwards of 4000 experiments on each network. Statistically this should not make the results differ, and indeed the same experimental upper boundary was obtained. But our method makes it easier to inspect the structure of each experimental network and understand how its structure relates to its tendency to exhibit TNCs.

These thoughts lead us to conclude that Scenarios 2 and 3 in finite networks are different phenomena near or above the theoretical upper boundary and have different enabling conditions. Indeed, while Scenario 2 requires only that the seed link to some vulnerable clusters, Scenario 3 requires that the flipping process "escape" the first vulnerable clusters linked to by the seed. This in turn requires that one or more stable nodes be flipped and that these flipped stable nodes also link to new vulnerable clusters not linked to by the seed so that these new vulnerable clusters will flip. The combined network of stable and vulnerable nodes that must be analyzed in order to understand this escape process is highly cross-linked. In fact, every stable node that flips is part of a closed loop, and the first few stable nodes that flip are part of closed loops that are quite short, containing perhaps as few as 5 nodes. But percolation theory as defined by Calloway et al and Watts works only on networks that are tree-like. Thus we conclude that percolation theory as applied by these authors is not able to describe the process that results in TNCs near or above the theoretical upper boundary in finite networks (although an analytical explanation could be possible). The remainder of this paper is devoted to using numerical experiments to understand this escape process in finite networks of various sizes above the value of z at which infinite networks display their upper boundary. No new explanation is necessary for what happens well inside the boundary because the vulnerable cluster is so large, as shown on the left in Figure 2.

In each experiment, data were recorded on how many vulnerable clusters the network had, the degree of the seed, how big the largest vulnerable cluster was, which clusters of vulnerable nodes were participating in the cascade, which vulnerable clusters were linked to by the seed, how large each participating vulnerable cluster was, when each stable and vulnerable node flipped, when a new vulnerable cluster joined the cascade, and how many vulnerable and stable nodes flipped at each stage of the cascade. These experiments were performed for $K^* = 8$, for which the experimental boundary occurs at $z \approx 11.5$, and $K^* = 5$, for which the experimental boundary occurs at $z \approx 6.5$.[5] Most of the numerical results discussed pertain to $K^* = 8$.

IV. Characteristics of Experimental Results

The experiments show several things. First, TNCs become rare at values of z that correspond to Watts' experimentally-determined upper boundary (the black dots in Figure 1). Cascades corresponding to Scenario 2 are not rare at these z values but instead become rare at much larger values. Second, TNCs can start when a seed links to only a few vulnerable nodes that lie in small clusters. Many, but not all, TNCs start when the seed links to more than two vulnerable clusters, a rare event. It is not necessary, nor is it often observed, that the seed link to a large or the largest vulnerable cluster. Instead, large clusters are often brought into the cascade at later steps. However, a TNC never starts if the seed links only to vulnerable nodes in clusters of size = 1, meaning that about 45% of the vulnerable nodes are unable to launch TNCs. Furthermore, only rarely is the largest vulnerable cluster in a network able to start a TNC if the seed links only to it. Finally, TNCs do not proceed by flipping all or even very many of the vulnerable nodes first and then turning to the stable nodes. On the contrary, TNCs proceed by almost immediately starting to flip stable nodes. In fact, if this does not happen within the first few steps, a TNC does not occur.

---

[5] For networks with fewer nodes, successively slightly larger values of z define the boundary. This is part of the size effect of using finite networks, discussed more later in the paper.

Two properties of the finite networks were observed to favor TNCs. One is the occurrence of some particular arrangements of vulnerable and stable nodes, called motifs below. The other is the presence of vulnerable clusters that are larger than occur on average in networks of that size and z. Without the motifs, TNCs cannot start. Without larger than average vulnerable clusters, TNCs are unlikely to survive beyond the starting stage enabled by the motifs. Each of these factors is in turn enhanced by smaller z, indicating that they are correlated. Furthermore, these factors differ greatly between different nominally identical networks (i.e., networks having the same size and z). Also, as the networks increase in size, z must be smaller in order for TNCs to occur at the same rate. Evidently, motifs become fewer and vulnerable clusters become relatively smaller as network size increases, and these effects must be compensated by decreasing z. The following sections provide detail on these findings.

V. The Cluster Seed Motif

In this section we discuss the first factor, two main patterns of node interconnections that enable TNCs to start and without which TNCs cannot start. Both of these provide the necessary conditions for flipping a first-order stable node: two flipped neighbors.

Here is what is observed when a TNC begins: The seed links to and flips one or a few vulnerable nodes, usually in more than one cluster. Within the first few steps, a flipped pair of vulnerable nodes has a common first-order-stable neighbor;[6] this neighbor flips. In a few of these instances, this neighbor has a vulnerable neighbor in another cluster. This vulnerable neighbor then flips, carrying the cascade to another group of vulnerable nodes, enabling the necessary cluster-hopping. This arrangement of nodes is here called a motif in the spirit of [Milo et al]. The first of two common

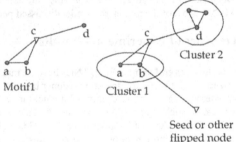

motifs is illustrated in
Figure 3. The first 10 or 20 steps of a TNC typically proceed by repeated cluster-hopping events enabled by these motifs. Only after this does the cascade proceed by stable nodes flipping each other.

---

[6] It is possible that this stable node is not first-order stable, but the likelihood of this event is about 0.1, so it is ignored here.

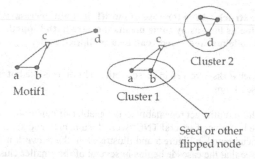

Figure 3. Illustrating the First Motif. Closed circles are vulnerable nodes while inverted triangles are stable nodes. (Left) Motif1 comprises a pair of vulnerable nodes *a* and *b* in the same or different clusters, their common first order stable neighbor *c*, and that neighbor's distinct vulnerable neighbor *d* in a different cluster from that or those of the original vulnerable pair. Note: If nodes *a* and *b* are in different clusters but by some means are both flipped, then the motif does not require these nodes to be connected. Such cases occur frequently but are not illustrated here. (Right) Example of motif1 in operation: The seed or another flipped node links to and flips vulnerable node *b* in Cluster 1. On the next step, vulnerable node *a* flips. This pair can then flip the first-order-stable node *c*, which in turn flips vulnerable node *d* in Cluster 2. In this way the cascade "hops" from Cluster 1 to Cluster 2.

Another motif, a subset of motif1, is shown in Figure 4 and is called motif2. Some of the ways it has been observed to act are also shown. Rarely, other patterns enabling TNCs to launch are observed, but the ones in

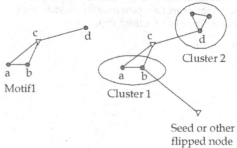

Figure 3 and Figure 4 are the most frequent.

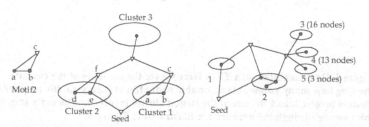

Figure 4. Illustrating the Second Motif. Left: motif2. Center: A documented start of a TNC which depends on motif2. Cluster 1 has 4 nodes while cluster 2 has 7. Only those involved in the TNC are shown. Right: Another documented start. Cluster 1 has one node while cluster 2 has 3. The TNC breaks out when

clusters 3 – 5 are struck. As in the case of motif1, it is not necessary that nodes *a* and *b* be connected as long as by some means they are both flipped. Also, as with motif1, the node marked "seed" can be any flipped node.

Note that the subnetworks involved in these motifs are not tree-like but instead contain short closed loops.

To determine if these motifs act repeatably as the enabler of cluster-hopping at the start of TNCs, the histories of several TNCs were documented and studied in detail. An example is documented in Figure 5 and illustrated on the network in Figure 6. In Figure 5 we can see that the cascade begins in several of the smaller clusters. Stable nodes are flipped almost right away, and almost every time another stable node is flipped, another cluster joins the cascade on the next step. By the 12th step, when only 145 out of 552 vulnerable nodes have been flipped, 155 stable nodes have been flipped. In these experiments, as soon as the number of stable nodes flipped at one step exceeded the number of vulnerable nodes flipped at that step, a TNC always occurred. The hypothesized mechanisms are clearly visible: the cascade proceeds initially by recruiting more and more relatively small clusters (the largest in this network contains only 13 nodes) and still more stable nodes until the cascade can be carried by the stable nodes flipping each other. Typically less than a dozen steps are needed in this region of the space to establish propagation via stable nodes flipping each other.

The early stages of this cascade through the network are documented in Figure 6. In this figure, only the local region around the first few participating nodes is shown, and only those links along which the cascade proceeds are shown. Since $z = 11.62$, each node actually has many more links than are shown. Several instances of cluster-hopping via these motifs can be seen. Other, more rare, patterns also are seen, such as the seed linking to a stable node that is flipped on the third step.

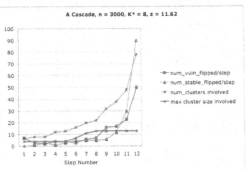

Figure 5. Documentation of a TNC. Here we see the statistics of the cascade, showing how many vulnerable and stable nodes flip at each step, the number of clusters involved, and the size of the largest cluster involved. The early stages of this cascade through the network are illustrated in Figure 6.

292

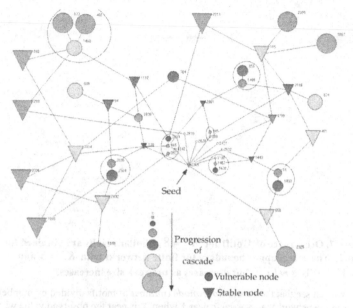

Figure 6. The First Seven Steps of the TNC Charted in Figure 5. Successive steps in the cascade are marked by nodes of increasing size. Vulnerable clusters are surrounded by ellipses. The seed is node 2365. Motif1 is the means by which stable nodes 1443, 2861, 1172, 2189, 94, and 2692 carry the cascade to new vulnerable clusters. The smallest nodes (199, 1198, 1342, 2427, 2532, 2678, and 2839) are struck by the seed at step 1. Node 2678 plays no further role. At step 2, vulnerable nodes 165, 382, 1457, and 345 flip. On step 3, 2365 and 165 flip stable neighbor 139, while 165 and 345 flip stable node 2861, and 2532 and 1457 flip stable neighbor 1443. On the next step, 139 carries the cascade to a new cluster, flipping 2605, while 2861 and 1443 do the same in two other new clusters. Node 1667 at the upper right, reached on step 7, is a member of a cluster of size 11. The largest cluster is reached on step 8. Many other flips can be seen that are not described verbally here. By the 6th step the cascade has a firm foothold and the motifs are no longer necessary to carry the cascade forward.

To determine the frequency of occurrence of the motifs illustrated in

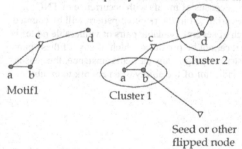

Figure 3, random networks with n ranging from 2000 to 18000 were generated, all pairs of vulnerable nodes were examined, and instances of the motif were recorded. The results are in Figure 7.

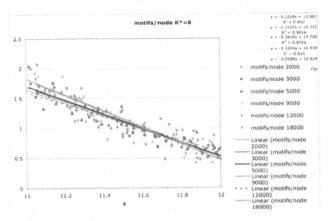

Figure 7. Occurrence of Motif1 for $K^* = 8$. **Similar results are obtained for motif2. The actual upper boundary for finite networks with** $K^* = 8$ **and** $n = 10000$ **is** $z \approx 11.5$, **but decreases as network size increases.**

Here we can see that the metric *motifs/node* (number of motifs divided by number of nodes in the network) takes a value near 1 when z is near the observed value where TNCs are rare. As noted above, smaller z is needed for larger n. The metric takes a value near 1 when $n = 10000$, but the observed value of the metric has no particular meaning and is used only for comparison between networks with different K^* to see if the pattern is repeatable.

Normalizing the occurrence of motifs by the size of the network is reasonable. One might think that, since the right lucky pair of vulnerable nodes must be found in order to activate either motif, one should normalize by the number of possible pairs of vulnerable nodes. However, it is not necessary for the seed to link to the lucky pairs but only that it link to vulnerable clusters containing the lucky pairs. The number of vulnerable clusters scales approximately linearly with the size of the network, so network size is a reasonable choice for normalizing factor.

We verified that absence of motif2 (and usually motif1 as well) is sufficient to prevent a cascade from leaving the first vulnerable clusters linked to by the seed. An example is shown in Figure 8.

It should be noted that associating the census of motifs with occurrence of TNCs is fraught because a single set of nodes arranged in the required pattern will be counted more than once, depending on which of several available pairs of vulnerable nodes is chosen as the starting point. Since it cannot be predicted which, if any, of these pairs actually are in a position to aid cluster hopping in any particular instance, the census of motifs can at best be a statistical indicator of the ability of a network to exhibit TNCs.

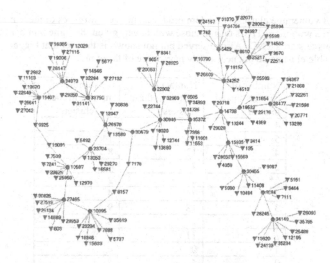

Figure 8. No Motifs: No TNC. A Large Vulnerable Cluster with No Occurrences of Motif1 or Motif2. Vulnerable nodes are closed circles while stable nodes are inverted triangles. All immediate neighbors of the 22 vulnerable nodes in this cluster are shown. The seed hit one of these vulnerable nodes but no TNC occurred.

TNCs start when several vulnerable clusters are linked to by the seed and when these clusters are of above average size, indicating that recruiting motifs is difficult and requires many vulnerable nodes to be involved. Table 1 provides some data.

Size of Network	Average size of first flipped clusters where TNCs occurred	Average number of first flipped clusters where TNCs occurred	Average size of first flipped clusters where no TNCs occurred	Average number of first flipped clusters where no TNCs occurred
4500	5	3	2	1.5
9000	9	3	3	1.5
18000	15	3	2.5	1.5

Table 1. Representative Statistics on Clusters Flipped by the Seed, Comparing TNC vs no TNC. Successful seeds link on average to twice as many vulnerable clusters as unsuccessful seeds do (compare column 3 to column 5), and those clusters are more than twice as big as those linked to by unsuccessful seeds (compare column 2 to column 4). The averages in the two right columns are essentially the network averages since TNCs occur less than 1% of the time. The largest vulnerable clusters in these networks are about 2 to 3 times larger than the averages shown in columns 2 and 4 respectively. Variability in these statistics is high since only a few clusters are flipped by seeds near the upper boundary.

TNCs are more likely when the seed strikes clusters that collectively are rich in motif1 and motif2 compared to clusters picked at random. Table 2 provides some representative data for 13 TNCs. In contrast to these data, individual vulnerable clusters selected at random have few or no occurrences of either motif. Also, picking pairs of vulnerable clusters at random produces more occurrences of these motifs in networks with more TNCs than in those with fewer, given the same number of attempts to start a TNC. Also observed but not shown is the fact that bigger vulnerable clusters do not necessarily have more of these motifs.

Number of clusters linked to seed	Number of motif1 in these clusters	Number of motif2 in these clusters
4	14	9
5	36	14
4	13	8
2	7	4
2*	0	2
2	7	4
3	18	9
2	2	4
3	11	6
1**	0	0
2	5	2
3	7	6
5	36	14

Table 2. Occurrence of Motif1 and Motif2 in Vulnerable Clusters Linked to by the Seed in 13 TNCs in Network test10000a, n = 10000, z = 11.417. This network has 1986 vulnerable nodes in 1167 clusters, the biggest cluster having only 21 nodes. *This TNC started via the middle mechanism in Figure 4. **This TNC started when the seed flipped a component containing a single node. This node and the seed linked to another stable node which then flipped. It flipped a vulnerable node in a larger cluster and the TNC proceeded from there via motif1.

Table 2 indicates that successful TNCs usually begin when the seed links to several clusters. This happens rarely. Note that while most of the clusters that launched the TNCs in Table 2 are relatively rich in the motifs, this is not necessary. One motif will do the trick. But to sustain the cascade it helps to have several so that the cascade can jump successively to new clusters. It also helps if the network contains relatively large vulnerable clusters.

VI. Size Effects

As the networks get bigger it is increasingly difficult to start TNCs given the same nominal z. Percolation theory says that this is because the size of the average vulnerable cluster is decreasing as a fraction of the network's size. Our experiments indicate that the size of the largest vulnerable cluster indeed falls as a fraction of network size given the same nominal z. But additionally, the seed needs to find a lucky combination of several clusters that are rich in motifs, and as the network grows, these are scattered among a growing number of ill-equipped clusters. The average seed has z links regardless of the size of the network and it hits on average about $0.15 z$ vulnerable nodes (typically in different clusters) when $z = 11.5$ so the search efficiency of seeds falls as network size grows. The importance of large vulnerable clusters is discussed again in the next Section .

VII. Effects of Variability

There is great variability in many of the statistics gathered during these experiments. For example, the size of the largest cluster in networks of the same size and practically same nominal z can vary by a factor of 2 in either direction from the mean. This strongly affects the rate at which TNCs occur, as illustrated in Table 3. In addition, as noted above, the census of motifs can only be regarded as indicative of TNCs and is also quite variable within and between networks with the same z. The census of motifs and the size of the largest vulnerable cluster are both correlated with z. A regression analysis was thus deemed appropriate to seek an understanding of the relative influences of these factors. This analysis reveals that z and *maxclust* (the size of the biggest vulnerable cluster) can predict $TNC/2000$ (the frequency of TNCs in 2000 tries) with $R^2 = 0.718$ and $p < 0.03$. A network with $n = 9000$, $z = 11.59$, and unusual (compared to Table 3) largest vulnerable cluster having 71 nodes launched a relatively huge 188 TNCs in 2000 trials.[7] Including the census of motif1 in the regression produced similar results but with smaller R^2 and p.

name	z	maxclust	TNC/2000
test9000e	11.4793	17	10
test9000d4	11.4938	36	6
test9000d4	11.4938	36	12
test9000d	11.4987	35	15
test9000d6	11.5064	31	10
test9000d5	11.5087	13	1
test9000bn	11.5322	22	4
test9000ak	11.5547	25	4
test9000at	11.5596	32	6
test9000aL	11.5711	21	1
test9000bm	11.5767	11	0

Table 3. Occurrence of TNCs in 2000 Trials in Networks with 9000 Nodes and Similar Nominal z and Different Size of Biggest Vulnerable Cluster

The occurrence of motif1 and motif2 also varies by a factor of 2 for different networks having the same z, as can be observed in Figure 7. When clusters are sampled randomly in the same network, a wide range of motif1 and motif2 can be found. These variations persist across networks even though the number of vulnerable nodes and vulnerable clusters each vary by much smaller percentage ranges.

Thus the conditions that favor respectively launching and sustaining TNCs are unevenly distributed between nominally similar networks and within each network. One may conclude that characterizations of cascade phenomena based only on first moments will not capture all of the effects or predict accurately when TNCs will occur in finite networks or predict which apparently similar networks can exhibit them.

VIII. Effects of Increasing z

---

[7] These are unique TNCs. Occasionally the same seed is used more than once. Such duplications are not counted.

As z increases for fixed n and K^*, first TNCs disappear and then Scenario 2 events disappear. As discussed above, TNCs disappear because motif2s disappear. Also, vulnerable clusters decrease in size, and the variability of their sizes, observed to be a factor, also decreases (measured by their coefficient of variation). The decrease in size and size variability of vulnerable clusters also extinguishes Scenario 2 events. For $K^* = 8$, Scenario 2 events disappear for $z \approx 17$ where all but a handful of vulnerable nodes are isolated. At this point about 1.3% of the nodes are still vulnerable.

IX. Experiments with Network Structure

A random model is suitable for analysis but real networks have some structure that might make cascades easier to launch. Accordingly, we studied two sources of structure, the degree correlation and the clustering coefficient.

Random networks with 3000 nodes were generated with various values of z corresponding to the upper boundary value, and the location of the boundary was confirmed using an appropriate value of ϕ. Once a statistical base was established at the boundary, the clustering coefficient or degree correlation of these networks were then raised by means of directed degree-preserving rewiring[8] [Maslov and Sneppen].

The rewired networks were then given randomly selected seeds in 100 trials and the occurrence of TNCs was noted. In both cases, the value of z at which TNCs occurred increased sharply as either r or c was increased, compared to rare occurrence in the base case. See Figure 9. Quite small increases in r or c were sufficient. The reason is not hard to find in the case of r. Only a slight increase in r dramatically increases the size of the largest vulnerable cluster: for $K^* = 8$ and $n = 3000$, the size of the largest vulnerable cluster is 19 when $r = 0$, 90 when $r = 0.1$ and 275 when $r = 0.2$. The value when $r = 0.2$ corresponds to over half the vulnerable nodes at the value of $z = 11.5$ that marks the original boundary when $r = 0$.

In the case of elevated c, no increase in the size of the largest vulnerable cluster, and no change in cluster size distribution, is observed. However, even a small clustering coefficient ($c = 0.0987$) generates enough triangles to enhance the ability of TNCs to start.

---

[8] A simple Matlab routine makes trial pairwise degree-preserving rewirings and accepts them if they yield a change in the target metric (r or c) in the desired direction. The metric can be moved in either direction but while positive or negative values of r can be obtained, only positive values of c can be. The networks are no longer random but they retain the degree sequence of the base network for comparison purposes.

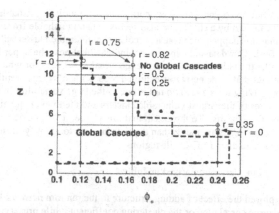

Figure 9. Diagram of Cascade Window (adapted from [Watts]) Showing effect of Elevated Degree Correlation. The dashed line window is given by theory. The solid circles represent simulations done by Watts using random networks with 10000 nodes. The open circles represent simulations done for this paper using random networks with 3000 nodes. Small differences between solid circles and open circles marked "$r = 0$" are not considered significant. Other open circles are the result of experiments with networks having $r > 0$. They represent combinations of z and r at which barely one cascade attempt in 100 succeeds, a working definition of the location of the upper boundary of the cascade window.

In both cases of enhanced structure, successful seeds are either of average nodal degree (enhanced c) or of smaller than average degree (enhanced r), whereas for no enhancement of structure, successful seeds are larger in degree than average.

X. Conclusions

We have studied the mechanism by which total network cascades (TNC), which consume an entire finite random network, can start in the region above the upper boundary derived by Watts for infinite networks. We distinguished TNCs from cascades that consume only the vulnerable clusters linked to by the seed (or possibly a few more), which are analogous to the cascades in infinite networks predicted by percolation theory. TNCs can start in the absence of large clusters of vulnerable nodes. The mechanism is different from that predicted by percolation theory, which is silent on the fate of stable nodes and speaks only to the behavior of vulnerable nodes. The mechanism is shown to be related to the prevalence of particular motifs of vulnerable and stable nodes by which the cascade is enabled from its first step to hop from one usually small vulnerable cluster to another, while at the same time starting to recruit stable nodes. If these motifs are absent, no TNC will occur. Thus a dispersed and apparently weak set of vulnerable nodes can "cause" a cascade if they can use stable nodes as bridges to link themselves to each other. The presence of relatively large vulnerable clusters is necessary for TNCs to grow beyond the first few vulnerable clusters, and the fall in relative size of large vulnerable clusters as n rises for the same z is the main size effect in reducing the likelihood of TNCs in networks with nominally the same z.

Watts noted that TNCs can start in networks that seem indistinguishable from their fellows and can start by a shock that also seems indistinguishable from others. He suggested that this happens because a percolating vulnerable cluster still exists, albeit difficult to find. Our data and experiments indicate that near the upper boundary in finite networks the vulnerable clusters are too small to account for what happens. Instead, the fact that some networks exhibit TNCs while other apparently identical ones (that is, having the same z) do not is due to the large variability in the enablers, namely the sizes of the largest vulnerable clusters and the uneven quantity and distribution of the motifs. To the extent that this is true, it will be necessary to use metrics based on higher moments than the first in order to identify these TNC-prone networks or locate their TNC-fertile regions.

The above findings are summarized in Table 4.

We also explored the effect of adding structure to the random network by increasing either the degree correlation or the clustering coefficient while preserving the degree sequence. Each modification makes TNCs easier to start. Increasing r even a little dramatically increases the size of the largest vulnerable cluster while increasing c a very modest amount adds many cross-links that make it much easier to flip stable nodes right at the beginning of the cascade. In each case, successful seeds are different in average degree from the base case, as are the respective mechanisms by which cascades carry forward after the seed acts.

	Finding	Finite Network	Infinite Network
Scenario 2	Cascade mechanism	Flipping the vulnerable clusters linked to by the seed	Same as for finite network: this kind of cascade satisfies the conditions set by percolation theory
	Max value of z ($K^* = 8$)	~17 (less than 1% of nodes are in clusters having more than one node)	~10.66
	Reason for extinction as z increases	No vulnerable clusters with more than one node	Vulnerable cluster size is finite
Scenario 3 (TNC)	Cascade mechanism	Cluster-hopping beyond the vulnerable clusters linked to by the seed	No theory yet, no experiments possible
	Max value of z ($K^* = 8$)	Corresponds to black dots; z~11.6 for $n =$ 10000; falls as n increases	No theory yet, no experiments possible
	Reason for extinction as z increases	Too few motif2, biggest vulnerable cluster too small, falling search efficiency of seed	No theory yet, no experiments possible

Table 4. Summary of Experimental Findings

XI. Future Research

Possible directions for future research can take two routes, depending on whether one views theoretical infinite networks as the canonical ones with explanatory power and finite networks as supporting or illustrative approximations of them, or (like the present paper) views finite (i.e., real) networks as the ones to study and the theory about infinite ones as a route to better understanding the behavior of finite networks. The former path suggests that additional attention to percolation theory may allow prediction of Scenario 3 events separately from Scenario 2 events in infinite networks as well as development of additional metrics for predicting when TNCs will occur in finite networks. The latter path suggests seeking to better understand the structural differences between real networks that do or do not exhibit TNCs, such as additional kinds of motifs or better identification, characterization, and counting of the motifs identified so far. In addition, either path could further study the deliberate insertion of non-random structure in order to better understand what enables or disables TNCs.

XII. Acknowledgements

The author thanks Duncan Watts for carefully reading an earlier draft, for many helpful and valuable communications and explanations, and for access to his paper with Peter Dodds "Influentials, Networks, and Public Opinion Formation," of which an advance version is available online at http://www.journals.uchicago.edu/JCR/. The author also thanks David Alderson, who also read an earlier draft and provided valuable feedback.

XIII. References

Calloway et al *Phys. Rev. Lett.* **85** *(25)* 5468-5471
Maslov and Sneppen *Science* **296** 910 3 May 2002
Milo et al *Science* **298** 824 25 Oct 2002
Watts *PNAS* **99** 5766 30 April 2002
Watts b (personal communications)

Index of authors

Printed in the United States
By Bookmasters